Report 128   1994

Guide to the design of thrust blocks for buried pressure pipelines

A R D Thorley and J H Atkinson

CIRIA

CONSTRUCTION INDUSTRY RESEARCH AND INFORMATION ASSOCIATION
6 Storey's Gate, Westminster, London SW1P 3AU
Tel 071-222 8891   Fax 071-222 1708

Summary

This report presents a design guide for thrust blocks to restrain the forces generated by changes in direction of fluid flow in jointed buried pressure pipeline networks. A step-by-step design approach is detailed. Refinements can be made by reference to detailed appendices, which give a background to the underlying principles and theory.

The guidance given in this report is principally for thrusts up to 1000 kN, limiting both the pressure range and pipe diameters and, more importantly, the thrust block sizes. While the theory used is applicable in any circumstances, the design of large scale thrust blocks requiring structural reinforcement, etc, is not covered.

Thorley, A.R.D. and Atkinson, J.H.
Guide to the design of thrust blocks for buried pressure pipelines
Construction Industry Research and Information Association
Report 128, 1994

© CIRIA 1994

CIRIA ISBN 0 86017 359 3

Thomas Telford ISBN 0 7277 1975 0

ISSN 0305 408X

Keywords (from Construction Industry Thesaurus)	Reader Interest	Classification	
Pipelines, thrust blocks, subsurface, water supply	Consulting engineers, contractors, water supply industry	AVAILABILITY CONTENT STATUS USER	Unrestricted Advice Committee guided Engineers, designers, contractors

Foreword

This report presents the results of a study to understand how thrust blocks act in the ground to restrain applied fluid forces in jointed buried pressure pipelines and to provide a design guide for the majority of cases (up to 1000 kN applied thrust).

The work was undertaken by Professor A R D Thorley and Professor J H Atkinson of City University under contract to CIRIA.

The research project was coordinated and managed by CIRIA with the assistance of a Steering Group who advised on the content of the work and the validity of the results. The Steering Group, which provided a balanced representation of the different interests involved, comprised:

Mr J Collins	A H Ball
Mr J Crossley	Yorkshire Water
Mr R D Currie	Johnston Pipes Ltd
Mr D Duffin	Severn Trent Water
Mr J D Hendry	Alfred McAlpine
Mr G Gray	CIRIA
Mr C B Greatorex	Stanton Plc
Dr S T Johnson (Chairman)	CIRIA
Mr D J Mackay	North West Water
Mr J Oliffe	Montgomery Watson
Mr J Stables	Wessex Water
Mr W R Waller	Acer
Mr C Waters	Thames Water

Acknowledgements

This project was funded by CIRIA and the following organisations:

Johnston Pipes Ltd
Stanton Plc
North West Water
Severn Trent Water
Southern Water
Wessex Water
Yorkshire Water

CIRIA thanks the many individuals and organisations who provided information and data for this research project. CIRIA also acknowledges the positive collaboration of all those involved during the course of the project and during the final period for the design method.

Contents

Figures

Tables

Glossary

Thrust block   a block of concrete (sometimes reinforced) in contact with a pipe bend or other component to restrain movement due to internal hydraulic forces.

Pipeline components:

Air vent;
air valve   device fitted to pipelines for the controlled admission and/or release of air.

Anchored joint   joint in which the adjacent pipes are mechanically connected (e.g. bolted) together to avoid relative movement.

Bend,

     downturn   a bend in which the flow changes direction downwards in a vertical plane.

     horizontal   a bend in which the flow changes direction in a horizontal plane.

     upturn   a bend in which the flow changes direction upwards in a vertical plane.

Blank end   an end of a pipe length which is terminated by a blank flange.

Flow restriction   orifice or partly closed valve which restricts the flow of water through it.

Keyed raft   a concrete raft, with integral concrete projections into firm ground below, and onto which a thrust block is cast.

Puddle flange   a dummy flange, welded or cast onto the outside of a pipe which is then embedded in a restraining wall to prevent axial movement.

Reducer   tapering component to reduce a pipe diameter from a larger to smaller value.

Self-restraining joint   see Anchored joint.

Taper
(see Reducer)   tapering component to join two pipes of different diameters.

Tee-junction   a pipe junction in which a side branch leaves the main pipe line at an angle of 90°.

Wash-out   a drainage point, usually at a low section of a pipeline, to allow collected sediment to be flushed out and for draining the line for maintenance purposes.

Y−branches	a pipe junction at which main pipeline splits into two equal (or unequal) diameter branches

Pressures and forces in pipelines:

Design force, F_d	the force, based on the design pressure, that is used as the maximum force to be restrained by the thrust block and which arises from fluid pressures.
Design pressure	the maximum pressure that is assumed to occur in the pipeline after making allowances for normal and abnormal operating conditions and hydraulic testing.
Shock loading	impact loads, frequently associated with pipeline vibration, cavity collapse and check valve slam under transient flow conditions.
Steady pressure	pressures that are constant, or nearly so, with time.
Surge (pressure surge)	one of several terms used to describe flows and pressures that change with time. Usually associated with events such as loss of power to pump driving motors, rapid changes to valve settings, etc.
Test pressure	the pressure applied during a hydrostatic test of a (section of) pipeline to check for leaks and adequacy of any anchorage of the pipeline. Usually the maximum pressure the pipeline should ever experience.
Transient pressure (see Surge)	rapidly changing pressures in a pipeline under changing flow conditions e.g. pump starts, pump trips, etc.

Ground conditions:

Soil classification	a scheme for describing and classifying soil based on grain size, plasticity and cementing.
Grading	the distribution of grain sizes in a soil from clay sized (very small) through silt sized, sand sized up to gravel and larger sizes.
Fine grained soils	soils containing a high proportion of clay sized particles.
Coarse grained soils	soils with only a very small proportion of clay sized particles.
Mixed soils (also called well graded)	soils with a wide distribution of particle sizes.
Cemented soils	soils in which the grains are cemented together, usually by some kind of mineral deposit.

Drained soils (coarse grained soils)	soils in which pore water pressures remain hydrostatic.
Undrained soils (fine grained soils)	soils in which there is no drainage or volume change during loading.
Rock	materials which are relatively strong and stiff because the mineral grains are strongly cemented together.
Weathering	processes by which rocks are attacked by climatic agents (e.g. rain, wind, frost action) so they soften and weaken. Ultimately rocks weather completely to become soils.
Jointing and fissuring	discontinuities and cracks which occur in almost all rocks and in many soils. The presence of joints and fissures will weaken a rock.
Peat	a soil composed primarily of organic material derived from vegetation. Peats are usually very soft and weak.
Fill	soil which has been placed by man and usually compacted. The properties of fill depend on the nature of the soil and on how well it has been compacted.
Groundwater	water present in the pores of a soil.
Pore pressure	the pressure in the groundwater (applicable to saturated soils — i.e. soils in which the pore spaces are completely filled with water).
Water table	the elevation in the ground where the pore pressures are zero.

Ground properties:

Unit weight, γ (kN/m^3)	the weight of everything (soil grains and water) in a unit volume of soil or rock.
Stiffness	the response of a material to loading below failure (stiff = small strains, soft = large strains).
Strength	the ultimate shear stress that a material can sustain without failing (strong = large shear stress, weak = small shear stress).
Undrained strength, s_u	strength of a soil for undrained loading: depends primarily on the current water content.
Drained strength	strength of a soil for drained loading: depends primarily on the current normal effective stress and is given by $\tau' = c' + \sigma'_n \tan\phi'$, (see also cohesion).
Friction angle, ϕ'	component of drained strength.
Cohesion, c'	component of drained strength due to cementing.

Stress and pressure in the ground:

Normal stress, σ_n	stress acting normal to a plane.
Shear stress, τ	stress acting parallel to a plane.
Total stress, σ or τ	the full stress on the plane (i.e. the combined effects of the stresses in the soil and in the pore water).
Effective stress, σ' or τ'	the stresses on the plane from the soil grains only ($\sigma' = \sigma - u$, where u = pore pressure, and $\tau' = \tau$). Effective stresses control all aspects of soil behaviour such as strength and compressibility.
Earth pressure	horizontal stresses on the vertical faces of a thrust block.
Passive pressure	earth pressure on the face which is moving towards the soil.
Active pressure	the earth pressure on the face which is moving away from the soil.
Earth pressure coefficients, K_p and K_a	the ratio of the horizontal and vertical effective stresses for passive and active pressures.
Base shearing resistance	the horizontal shear stress between the ground and the base of a thrust block.
Bearing pressure, q	the vertical normal stress between the ground and the base of a thrust block.
Bearing capacity, q_b	the value of the bearing pressure at failure (i.e. the maximum bearing pressure that can be applied).
Bearing capacity factors, N_c, N_γ, N_q	factors relating the bearing capacity of a thrust block to the ground conditions.

Forces on a thrust block:

Ultimate resistance, R_u	the maximum force that can be applied by the ground to a thrust block: this is the sum of all the ground resistances and it applies for relatively large movements.
Nominal resistance, R_n	the force that can be applied to the smallest practicable thrust block based on nominal ground conditions.
Ultimate bearing resistance, Q_b	the maximum vertical force that can be applied to the base of a thrust block by the ground (i.e. $Q_b = q_b A_b$ where A_b is the base area).

Thrust reduction factor, T_r

the factor used to reduce the ultimate resistance so that the design force can be applied to a thrust block with relatively small ground movements.

Safety factor, F_s

the factor used to ensure that a thrust block has a satisfactory margin of safety against uplift due to upward vertical loading.

Notation

A_b base area of thrust block
A_f face area of thrust block
A_{fm} minimum area
A_{fu} nominal area
B width of thrust block
C minimum cover of concrete around pipe
C_c coefficient of compression (soil)
C_s coefficient of swelling (soil)
c' cohesion
c'_p peak cohesion
D_{max} maximum density of granular soil
D_{min} minimum density of granular soil
D_p external diameter of pipe
D_r relative density of granular of soil
F applied force
F_d design force
F_h horizontal component of thrust
F_s safety factor
F_u ultimate design force
F_v vertical component of thrust
k coefficient of permeability − in Darcys law
K_a earth pressure coefficient-active
K_p earth pressure coefficient-passive
N_γ bearing capacity factor
N_c bearing capacity factor
N_q bearing capacity factor
p pressure in pipe
q bearing pressure
q_b bearing capacity
Q_b ultimate bearing resistance
R_n nominal resistance (ground resistance to smallest practicable thrust block)
R_o overconsolidation ratio
R_u ultimate resistance (sum of all ground resistances)
s_u undrained strength
t time
T_r thrust reduction factor
u pore pressure
V_b volume of thrust block
v velocity of fluid in pipe
w water content of soil
W weight of thrust block
W_b weight of pipe bend
W_f weight of fluid
Z_b depth from ground surface to base of thrust block
Z_c depth from ground to centreline of pipe

γ unit weight of soil

γ_w unit weight of water

δ displacement

Δu change of pore pressure

θ angle of pipe bend

ρ density of fluid in pipe

σ total stress

σ_a normal stress on active face of thrust block

σ_b normal stress on base of thrust block

σ_n normal stress

σ_n' normal effective stress

σ_p normal stress on passive face of thrust block

σ_v total vertical stress

σ_v' vertical effective stress

τ shear stress (total)

τ' effective shear stress

τ_c' critical shear stress

τ_p' peak shear stress

ϕ' friction angle

ϕ_p' peak friction angle

1 Introduction

Fluid under pressure in any pipeline creates forces at bends, junctions, valves and all restrictions to, and changes in, direction of flow. Additional transient forces may be generated by pump starts or stops, valve closure, etc. During construction and commissioning pipelines are, or should be, subjected to hydraulic pressure tests which should be higher than the sum of the steady and transient pressures.

If the pipeline is continuous, or if it has anchored joints the forces may be resisted by tension and compression in the pipe, and by shear between the pipe and the surrounding soil. For pipes with joints which are not anchored, by welded or bolted flanges, etc., *thrust blocks* transmit these forces to the ground.

Thrust blocks normally consist of a volume of concrete, usually of nominal strength ($20-40$ N/mm^2), which may be lightly reinforced. The size and shape of the block is decided on the basis of the forces to be restrained, the size and style of the pipe fitting or component, and local ground conditions.

The effectiveness of any thrust block is determined by its mass, shape, position relative to the pipeline, the soil reactions on the block, and friction between the pipeline and the surrounding ground.

The design methods in this report are relatively simple and in most cases can be followed by non-specialist engineers. They are, of necessity, relatively conservative and, in general, they will tend to produce thrust blocks which are perhaps rather larger than would be obtained by using the services of hydraulic and geotechnical engineers.

If any user of this report believes that their design is over-conservative it is always open to them to return to first principles to seek a more economical design.

1.1 SCOPE

The report covers the routine design of thrust blocks for pressure pipelines which are:

- non-continuous and without anchored joints

- made from common materials (e.g. ductile iron, glass reinforced polyester (GRP), unplasticised polyvinylchloride (uPVC), asbestos cement and prestressed concrete)

- buried in trenches

- up to 1000 mm diameter, with test pressures up to 25 bar (2500 kPa) and/or with forces up to 1000 kN.

In other situations (e.g. larger sizes, higher pressures, etc.) the same general principles of hydraulics and soil mechanics apply but designs will need a strong element of engineering judgement and should be carried out, or at least reviewed, by engineers with the appropriate experience in hydraulics and geotechnics.

1.2 GENERAL PRINCIPLES

Fluid pressures in the pipeline will impose horizontal and vertical (upward or downward) forces at bends, valves, constrictions, junctions, etc. (see Figure 1). These forces must be resisted by the soil in front of and below the thrust block (see Figure 2) by a combination of:

• horizontal normal stress on the faces of the block

• vertical normal stress on the base of the block

• shear stress on the base of the block

• the weight of the block and soil above it.

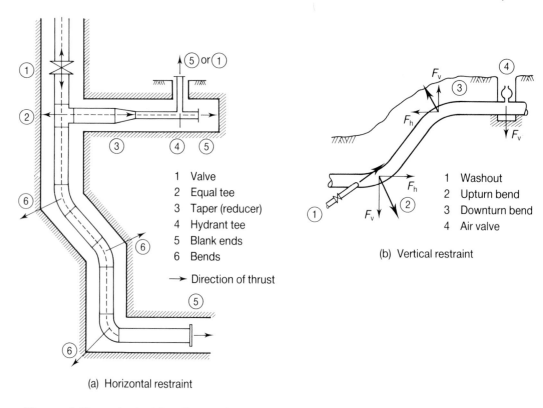

1 Valve
2 Equal tee
3 Taper (reducer)
4 Hydrant tee
5 Blank ends
6 Bends

→ Direction of thrust

1 Washout
2 Upturn bend
3 Downturn bend
4 Air valve

(b) Vertical restraint

(a) Horizontal restraint

Figure 1 *Some typical locations where restraint is necessary*

Thrust blocks must be designed to restrict movements so that pipe joints do not leak. For this reason the allowable soil stresses must be considerably smaller than those required to cause ultimate failure of the thrust block itself. For flexible jointed pipelines the consequence of failure is usually joint leakage followed by washout leading to further movement. Rupture is less likely to occur except where there is a gross failure of the restraint.

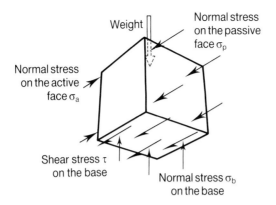

Figure 2 *Components of ground resistance on a thrust block (view from underneath)*

NOTE:

1. Other shear stresses acting on the sides and top of the block will be neglected.

2. Consideration must be given to the influence of groundwater in calculating soil stresses and weights.

3. The weight of soil above the thrust block may be included only if backfill is placed before the test pressure is applied.

4. The geometry of the block should be proportioned so that there are no excessive moments — normally the line of action of the imposed force should not fall outside the middle third of the bearing face.

The design procedure that has been adopted requires that the design force F_d due to fluid pressures must not exceed the ultimate resistance R_u of the thrust block, after this has been reduced by a Thrust Reduction Factor T_r, so that for a satisfactory design:

$$F_d < \frac{1}{T_r} R_u$$

The value of T_r will normally be in the range 2 to 5, depending primarily on the ground conditions.

NOTE:

1. The thrust reduction factor T_r contains an allowance for uncertainties in ground conditions and for approximations in the analyses. There is no need to apply an additional factor of safety.

2. The thrust reduction factor T_r is intended to ensure that the ground stresses remain within the range of relatively small strains to limit ground movement.

---

2 Design

2.1 CLASSIFICATION OF DESIGN

In any design process it is vital to recognise the magnitude of the problem, to adopt appropriate procedures and to employ an appropriate level of expertise. Two classes of thrust block design can be identified depending on:

1. The magnitude of the forces to be resisted, which are a function of pipe size, pressure and the type of component (bend, tee, etc.).

2. The ground conditions; the nature of the ground, its state or consistency (e.g. water content, relative density, fabric, weathering) and the groundwater conditions.

3. The consequences of failure; damage to the pipeline system, pollution, problematic repairs and the inconvenience and loss of revenue from interruptions to the service being provided.

The flow diagram in Figure 3 sets out the logical sequence for design and identifies two classes:

- Class 1 − Routine design

Design requires the use of simple routine calculations verified by on-site inspection and simple testing/assessment of ground conditions and is applicable to pipelines in good ground conditions (e.g. weathered rock, firm to stiff clay, cemented sands and gravels and engineered fill) **and** for small or low pressure pipes with forces less than 1000 kN.

- Class 2 − Detailed design/special considerations

Design requires a detailed assessment of the forces by engineers with experience in hydraulics and evaluation of the ground conditions by a geotechnical specialist and is applicable to pipelines in poor ground (e.g. soft clay, peat, loose sands and gravels, poorly compacted fill) or for large or high pressure pipes with forces greater than 1000 kN.

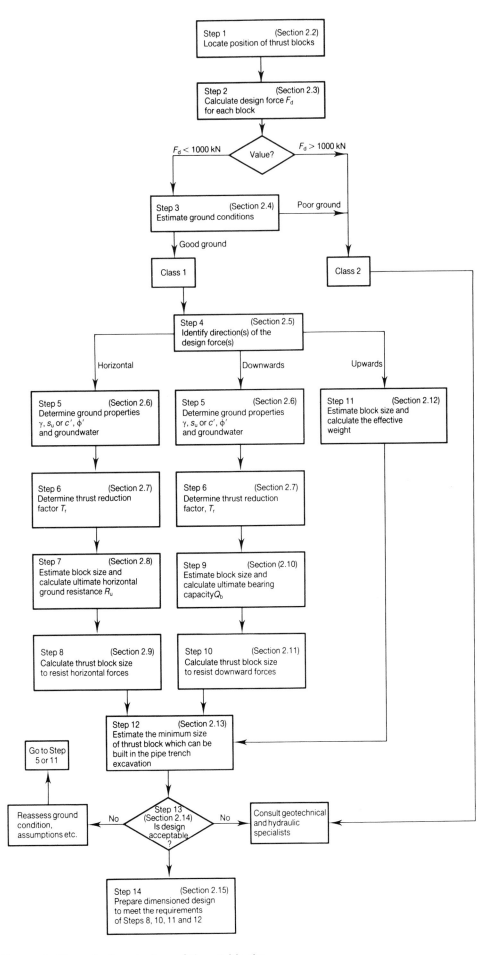

Figure 3 *Steps in the design of thrust blocks*

2.2 STEP 1 – LOCATE POSITION OF THRUST BLOCKS

Forces requiring restraint will be generated wherever there is a change in the direction of flow in either horizontal or vertical planes, and at bends, tees, valves, end stops, etc. Figures 4(a) and 4(b) illustrate, for a typical pipeline, where forces occur and the locations of thrust blocks to resist these forces.

NOTE:

1. There are a number of other positions where thrust blocks may be required although loads will not be generated when the line is in normal use, e.g. level invert tees, washouts and hydrant tees.

2. If a washout is not axially restrained relative to the tee branch line, then there will also be a reaction in normal pipeline use, and a thrust block will be required.

3. Consider re-routing the pipeline to avoid unnecessary changes in direction (e.g. at entries to, and exits from, pumping stations).

4. Consider the effect of any adjacent services or situations which may lead to subsequent excavation near to thrust block locations and so may undermine the capacity of the block.

5. Valves are often placed in chambers for reasons unrelated to the restraint of hydraulic forces although the chambers can provide an opportunity for restraint.

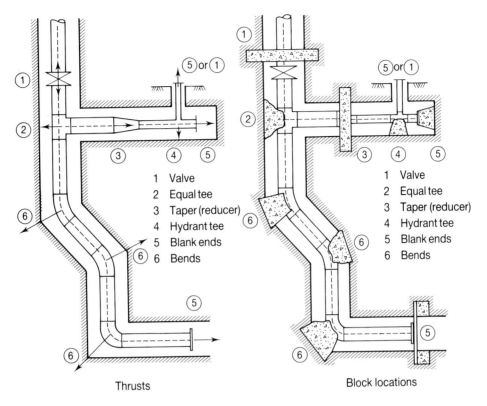

1 Valve
2 Equal tee
3 Taper (reducer)
4 Hydrant tee
5 Blank ends
6 Bends

1 Valve
2 Equal tee
3 Taper (reducer)
4 Hydrant tee
5 Blank ends
6 Bends

Thrusts

Block locations

(a) Horizontal thrusts

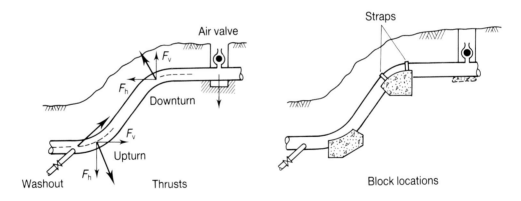

(b) Vertical thrusts (upwards and downwards)

Figure 4 *Typical locations for thrust blocks*

2.3 STEP 2 – CALCULATE THE MAGNITUDE OF THE DESIGN FORCE, F_d

The maximum thrust forces for design are those developed at the field test pressure, which is reached during commissioning, and exceptionally during operation. The test pressure value will be determined by the designer and in the UK the recommendations of BS 8010[1] will normally be followed. These state that the hydrostatic test pressure should be equal to the *working pressure × 1.5* or *working pressure + 5 bar* or the *maximum working pressure + surge pressure*, whichever is the greater.

In all cases the test pressure should not exceed the pressure rating of the pipe + 5 bar.

2.3.1 Influence of transients and surge pressures

During operation, peak pressures in the pipeline (those pressures above the working pressure but less than the hydrostatic test pressure) are usually associated with fluid transient or surge conditions, typically following pump trips or valve operations.

NOTE:

1. Peak pressures in a pump-fed main may be determined by the shut-off head of the pump when the line is shut off at the downstream end. If the discharge is at a lower elevation than the source the pressures may considerably exceed the pump shut-off head.

2. Rapid shock loadings under fault and transient conditions can be minimised by careful attention to choice of components, particularly check valves (to prevent check valve slam), air release valves and relief valves.

3. Conditions leading to cavitation must be recognised and where practical the design or specification should be changed accordingly.

2.3.2 Calculation of design force

Once the design pressure is established the design force for standard components can be obtained either from the formulae in Figure 5, from Table 1 or from Figure 6, whichever is most convenient. When using the tables and figures the 'effective diameter' is defined as the diameter over which the design pressure will actually be applied – usually this will equal the **external** diameter of the pipe.

Assumptions:

1. Flow cross-sectional areas are constant (except for tapers).

2. Viscous losses in components are negligible, except for partially closed valves.

3. The dynamic pressure head is small (flow velocities rarely exceed 3 ms⁻¹, equating to a dynamic head <0.05 bar).

4. One-dimensional flow theory is valid.

5. The liquid in the pipe is a Newtonian fluid, e.g. water.

6. Thrust forces from any friction generated as a result of a pressure drop associated with a change in flow velocity are considered negligible for applications within the scope of this report.

Component		Simple formula for hydraulic thrusts
Horizontal bend in the $x-y$ plane	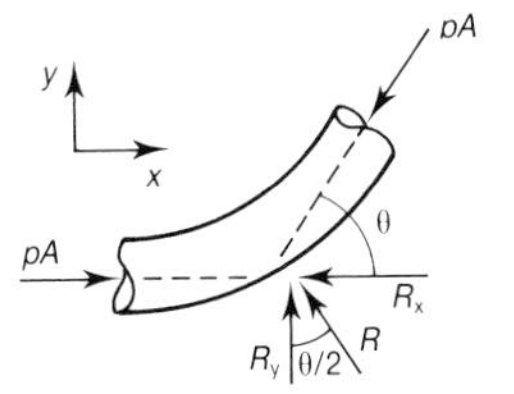	$R_x = pA(1 - \cos\theta)$ $R_y = pA\sin\theta$ Total thrust $R = 2pA\sin\frac{\theta}{2}$
Upturn in the $x-z$ plane	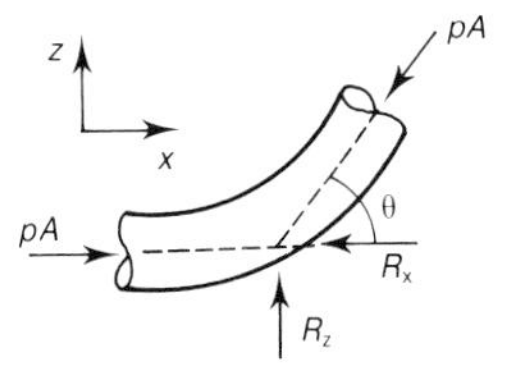	Horizontal component $R_x = pA(1 - \cos\theta)$ Vertical thrust $R_z = pA\sin\theta$
Downturn in the $x-z$ plane	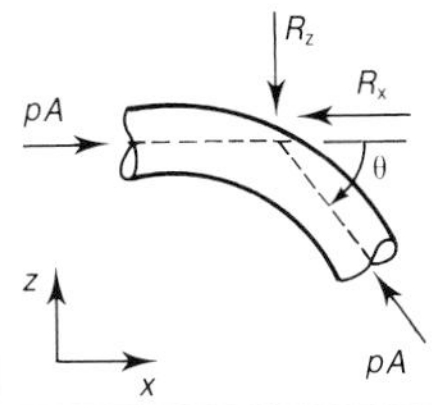	Horizontal component $R_x = p_1 A(1 - \cos\theta)$ Vertical component $R_z = pA\sin\theta$
Tee in the $x-y$ plane	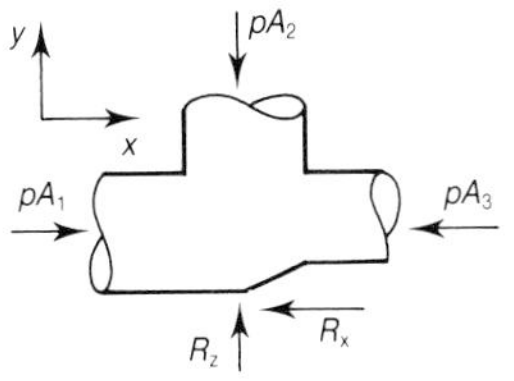	Axial thrust $R_x = p(A_1 - A_3)$ $= 0$ if $A_1 = A_3$ Side thrust $R_y = pA_2$
Taper/Reducer along the x-axis	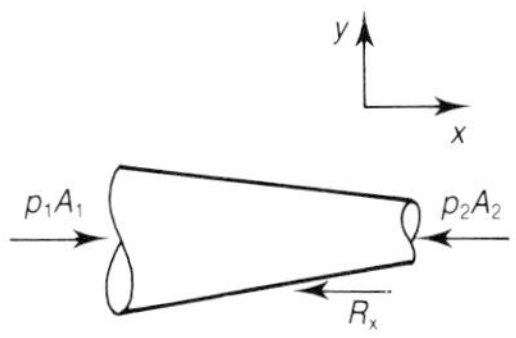	Axial thrust $R_x = p(A_1 - A_2)$
Dead end and closable valves	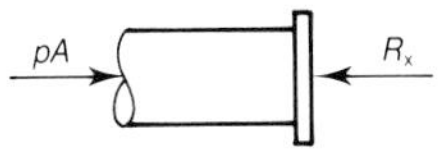	Axial thrust $R_x = pA$

Note: The resultant forces, and their components, are those due to hydraulic pressure only. It is assumed that the net weight of the thrust block and pipe are equivalent to the ground material that they are replacing.

In the above equations:

p = design pressure
A = pipeline cross-sectional area (based on the external diameter)
θ = angle through which the pipe bends

Figure 5 *Formulae for determining components of the hydraulic thrust forces requiring restraint in various fittings and components*

Table 1 *Loads (in kN) on blank ends and bends due to internal pressures of 10 bar*

Effective diameter (mm)	Blank ends and tees	Standard elbows and bends			
		90°	45°	22.5°	11.25°
50	1.96	2.78	1.50	0.77	0.38
75	4.42	6.25	3.38	1.72	0.87
100	7.85	11.1	6.01	3.06	1.54
150	17.7	25.0	13.5	6.89	3.46
200	31.4	44.4	24.0	12.3	6.16
250	49.1	69.4	37.6	19.1	9.62
300	70.1	99.9	54.1	27.6	13.9
350	96.2	136	73.6	37.5	18.0
400	126	178	96.2	49.0	24.6
450	159	225	122	62.0	31.2
500	196	278	150	76.6	38.5
600	283	400	216	110	55.4
700	385	544	294	150	75.4
800	503	711	385	196	98.5
900	636	900	487	248	125
1000	785	1111	601	306	154
1100	950	1344	727	371	186
1200	1131	1599	865	441	222
1400	1539	2177	1178	601	302
1600	2011	2843	1539	784	394
1800	2545	3598	1947	993	499
2000	3142	4442	2402	1226	616
2200	3801	5375	2909	1483	745
2400	4524	6397	3462	1765	887
2600	5309	7507	4063	2071	1041

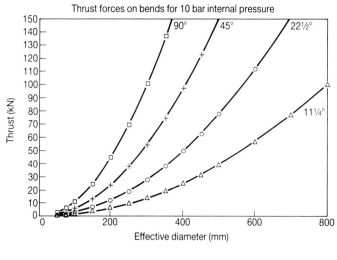

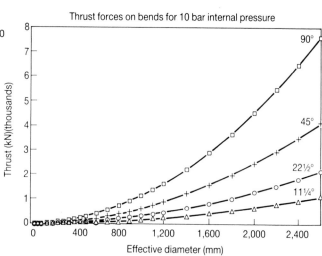

Figure 6 *Thrust forces on bends*

2.4 STEP 3 – ESTIMATE GROUND CONDITIONS

Ideally, ground conditions should be assessed before detailed design and construction by boreholes or trial pits excavated at, or near to, the location of each thrust block. In practice this ideal is rarely achieved and is often impractical and uneconomic, particularly for smaller diameter or low pressure pipelines.

In the absence of a site investigation, ground conditions can be assessed, directly and indirectly, by using information from:

- adjacent sites

- previous jobs in the locality

- geological and engineering geology maps and borehole information available from the British Geological Survey (BGS)

- walking the proposed route and noting outcrops and other geomorphological indicators

- samples taken from the trench walls during the pipeline construction.

Appendix 2 gives guidance on field identification of soils and rocks.

In the first instance, and in the absence of definite data, Table 2 can be used to estimate the ground conditions.

Table 2 *Estimating ground conditions*

Good ground:	Fresh and weathered rock
	Firm and stiff clay
	Cemented sand and gravel
	Dense sand and gravel above water table
	Engineered fill
Poor ground:	Soft clay
	Loose sand and gravel above water table
	Uncemented sand and gravel below water table
	Silts
	Peat
	Poorly compacted fill

2.5 STEP 4 – IDENTIFY DIRECTION OF THE DESIGN FORCES

Before proceeding any further it is necesary to identify the direction of the forces in order correctly to resolve the resultant force and direction.

The directions of the design forces for various components are shown in Figure 4.

2.6 STEP 5 – DETERMINE GROUND PROPERTIES

Before starting detailed design or construction, the ground conditions should be assessed at each thrust block location from whatever knowledge can be obtained from previous or current work in the area. This assessment should be confirmed by examination during construction. Boreholes or trial pits excavated at the location of each thrust block are the ideal but are often unavailable. The best opportunity for identifying the ground conditions generally occurs during construction of the pipeline when the pipe trench is excavated.

Many of the requirements for identifying the ground for design of thrust blocks are the same as those required for design of trenches described in CIRIA Report 97[2]. Additional guidance is given in Appendix 2 and further information can be obtained from many of the standard texts on soil mechanics and site investigation.

Ideally examination and assessment will be carried out by a geotechnical specialist who will identify:

- the type of soil or rock by grain size, and mineral composition

- the consistency, degree of compaction and degree of cementing

- groundwater conditions.

It is most important to distinguish soil which will behave in a *drained* manner from soil which will behave, essentially, in an *undrained* manner.

Drained soils

These are relatively coarse grained gravels, sands and silts. In these soils the permeability is relatively high and pore pressures will remain hydrostatic under working conditions.

For drained soils the relevant strength parameters are: friction angle, ϕ', and cohesion, c'.

In most cases the cohesion, c', will be taken as zero. Only for cemented coarse grained soils will values of c' greater than zero be used in the design of thrust blocks.

Undrained soils

These are the fine-grained clay soils, with a relatively low value for permeability and therefore changes of water content occur only very slowly.

In undrained soils the relevant strength parameter is the undrained shear strength s_u.

The undrained shear strength, s_u, of soil depends primarily on its water content so it is essential to determine undrained strengths at water contents corresponding to working conditions.

NOTE:

1. Drained soils are often called granular or frictional soils whilst undrained soils are often called cohesive soils. These definitions can be misleading as granular soils may be undrained (as in a sand castle) and clay soils may be drained (as in many coastal cliffs). In engineering practice it is better to classify soils according to their behaviour under the particular circumstances of the problem under consideration.

2. For drained soils the position of the groundwater table has a major influence on the size of the thrust block. If, for instance, the water table rises from below the base of a thrust block to above the pipeline, the capacity of the block will be approximately halved.

3. For undrained soils the position of the groundwater table does not significantly influence the capacity of the thrust block.

The ground strength may be assessed in a number of ways:

- from standard, routine laboratory tests (e.g. shear box and triaxial tests)
- from standard, routine, in-situ tests (e.g. SPT and vane tests)
- from observation of the type of soil and its consistency
- from specialised in-situ tests (e.g. cone penetrometer).

2.6.1 Strength

Table 3 summarises the principal soil types, their ranges of consistency, and approximate strengths. Used in conjunction with Appendix 2, initial values can be obtained for design purposes subject to field confirmation.

2.6.2 Unit weight, γ

This is the weight of soil and water per unit volume (γ has units of kN/m^3). A number of routine field tests are available for measuring the value of unit weight in-situ and in undisturbed tube samples. For routine design purposes it is sufficient to take the values given in Table 4.

Table 4 *Assessment of unit weight, γ,*

Soil/rock type	Unit weight, γ, kN/m^3
Clay, rock	20
Granular soil − below water table	18-20
Granular soil − above water table	16-18
Fill	16-20

2.6.3 Groundwater conditions

The condition for undrained soils (i.e. clays) is that their water content remains essentially unchanged during loading. They may be saturated or unsaturated and pore pressures may be positive or negative due to suctions developed by desiccation.

Table 3 *Description of typical soil properties*

Soil type	Category	Field identification	Consistency	Strength
Coarse grained soils (over 65% sand and gravel)	Boulders > 200 mm Cobbles 60−200 mm Gravel 2−60 mm Sand 0.06−2 mm Silt 0.002−0.06 mm	i. Particles visible with naked eye; except silt which is visible with a hand lens	Loose: can be excavated with a spade: 50 mm wooden peg can be easily driven	$\phi' = 30°$ $c' = 0$
		ii. Feels gritty	Dense: requires pick for excavation: 50 mm wooden peg hard to drive	$\phi' = 37°$ $c' = 0$
		iii. Forms a cone when dry or submerged		
		iv. Dry silt 'dusts off' hands or boots	Cemented: pick removes lumps which may be abraded	
			lightly	$c' = 10$ kPa
			moderately	$c' = 40$ kPa
			strongly	$c' = 100$ kPa
			(see Appendix A)	
Fine grained soils (over 35% clay)	Clay	i. Clay particles not visible	Very soft, extrudes between fingers when squeezed	s_u <20 kPa
		ii Clay can be moulded like plasticine	Soft: easily moulded in fingers	$s_u = 20 − 40$ kPa
		iii. Clay will not 'dust-off' hands or boots and must be washed or scraped off	Firm: can be moulded in fingers strong pressure	$s_u = 50 − 75$ kPa
			Stiff: cannot be moulded in fingers	$s_u = 100 − 150$ kPa
			Hard: brittle or very tough	$s_u > 150$ kPa
Organic soils	Peats	i. Dark colour		Usually unsuitable for thrust blocks
		ii. Often fibrous		
		iii. Characteristic odour		
Fill	All types	i. Note soil type	Well compacted	Assess strength depending on grain size and consistency as above
		ii. Note unnatural material	Uncompacted or poorly compacted	Usually unsuitable for thrust blocks
Rock	Many types	i. Usually stronger and stiffer than soils	Unweathered with no joints at 45°	$s_u > 150$ kPa $\phi' = 30°$
		ii. Note structure and orientation of joints and fissures	Unweathered with joints at about 45°	
				$\phi' = 30°$
			Many joints at many orientations	
			Weathered: assess as for soil	

Notes:
1. Soft clay deposits are often stronger close to the ground surface due to drying. However, the clay crust is usually cracked and fissured
2. Stiff clays will swell and soften on exposure to water.

Construction of trenches and thrust blocks will alter the stresses in the surrounding ground and this may lead to increases in the water content. Hence, for clays it is important to estimate the undrained strength at a water content which will be reached after construction and after swelling and consolidation to the long term equilibrium condition.

For drained soils, pore pressures remain hydrostatic and it is necessary to specify the worst conditions for design. If it is certain that the soil will remain dry, or unsaturated, with the water table *always* below the base of the thrust block then the soil conditions correspond to *above the water table* with $\gamma_w = 0$ in the design calculations.

If there is a possibility that the groundwater table will rise above the base of the thrust block it should be assumed that the water table may rise to ground level and then the soil conditions correspond to *below the water table*.

Intermediate conditions, where the water table is below ground level but above the base of the thrust block are not covered specifically in this report. The assumption that the water table is at the ground surface will lead to a safe solution for all other groundwater conditions. Alternative designs for intermediate groundwater table conditions should be carried out by geotechnical specialists.

NOTE:

1. Backfilled trenches often act as drains and the water table in the backfill is likely to be high, especially during and immediately after heavy rainfall.

2. For undrained (i.e. clay) soils there is no need to assess the position of the water table.

3. For drained (i.e. sands and gravels) soils assume the water table to be at ground level unless it can be demonstrated that for the life of the pipeline it will always be lower.

2.7 STEP 6 – DETERMINE THRUST REDUCTION FACTOR T_r

Increasing any force, F, applied to a thrust block (see Figure 7(a)) will cause displacement of the block. The relationship between load and displacement will be similar to that in Figure 7(b). When the ultimate ground resistance, R_u, is reached displacements become very large as the soil yields and eventually fails. At some smaller load, the displacements will be correspondingly smaller. The thrust reduction factor, T_r, is defined as:

$$T_r = \frac{R_u}{F}$$

where R_u is the ultimate ground resistance, and F is the force applied to the thrust block. The thrust block design is satisfactory if the design force, F_d, is less than the ultimate resistance R_u, reduced by a suitable reduction factor which will ensure that the displacements will be relatively small.

Values for thrust reduction factors for Class 1 thrust blocks are given in Table 5 for different soil and rock types. If these lead to unacceptably large thrust blocks, the reduction factor may be re-assessed by determining the actual relationship between thrust reduction factor and displacement under defined load and ground conditions.

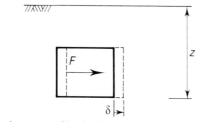

(a) Displacement of loaded thrust block

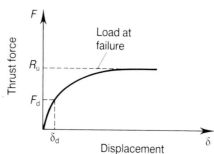

(b) Idealised load/displacement curve for a loaded thrust block

Figure 7 *Displacement of a thrust block under load*

The displacement of a thrust block will depend on:

• the size of the thrust block

• the applied loads

• the soil or rock type.

The size of the thrust block may be taken into account by relating the displacement, δ, to the depth, Z.

Table 5 *Reduction factors (T$_r$) for Class 1 thrust blocks*

Soil or rock type	Reduction Factor (T$_r$)
Dense sand or gravel Moderately or strongly cemented soil Stiff clay Fresh rock	2 to 3
Medium dense sand or gravel Lightly cemented soil Firm clay Weathered rock	3 to 4
Loose sand or gravel Soft clay	4 to 5

The loads may be taken into account by considering the thrust reduction factor.

The soil or rock may be either *soft* (e.g. loose sands and gravels soft clays) or *stiff* (e.g. dense sands and gravels, cemented soils, rocks).

The scale for δ/Z depends on the geometry
of the thrust block and on the soil or rock
and groundwater conditions. In many cases
the ultimate resistance will be reached after
displacements of the order of δ/Z = 10% to 20%.

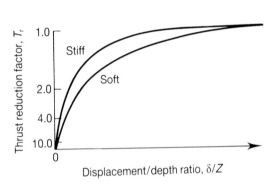

Figure 8 *Characteristic variation of thrust reduction factor T$_r$ with displacement ratio δ/Z for thrust blocks in stiff and soft soils*

Figure 8 illustrates typical curves of thrust reduction factor, T_r against displacement ratio, δ/Z for typical stiff and soft soils. For stiff soils displacements are considerably smaller than the corresponding displacements for soft soils at all stages so that, for a given displacement, larger reduction factors should be applied for soft soils and smaller reduction factors for stiff soils.

Design curves giving the relationships between load and displacement can be obtained by carrying out special field or laboratory tests and used to re-assess the value of the load factor required to limit the displacement of the thrust block to a value selected by the designer.

These analyses should be done under the direction of a geotechnical specialist. However, in order to ensure that the loading does not approach the ultimate failure state the load factor should not be less than 2 under any circumstance.

NOTE:

1. The thrust reduction factors in Table 5 are intended to include a safety factor to account for uncertainties in the calculation of fluid forces and ground resistance.

2. The reduction factors in Table 5 account for the different stress − strain and strength characteristics of different soils and rocks. They are intended to ensure that the ground strains remain within the range of relatively small strains necessary to limit ground movements.

3. If the reduction factors given in Table 5 result in unacceptably large thrust blocks then a number of factors may be reconsidered (e.g. ground conditions, design force, etc.).

4. For a given soil higher or lower values of reduction factor reflect the quality of the available soils information and the consequences of larger or smaller thrust block movements.

2.8 STEP 7 – ESTIMATE BLOCK SIZE FOR HORIZONTAL BENDS AND CALCULATE ULTIMATE HORIZONTAL GROUND RESISTANCE R_u

It is a basic requirement that the fluid thrust must not exceed the ground resistance reduced by a suitable factor to limit ground movements.

Horizontal ground resistance on a thrust block is the result of:

• active and passive earth pressures acting on the front and back faces

• shear stresses acting on the base.

As with all calculations in soil mechanics it is necessary to distinguish between drained and undrained conditions.

Derivations for the following equations are given in Appendix 4.

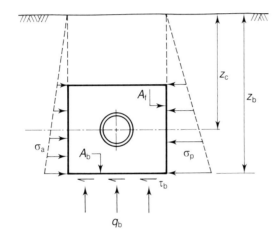

Figure 9 *Active and passive earth pressures on the faces of a thrust block*

The ultimate resistance to horizontal forces from passive and active pressures on the front and back faces of the block and the shear stresses on the base (see Figure 9) is:

$$R_u = \left(\sigma_p - \sigma_a\right) A_f + \tau_b A_b$$

where $(\sigma_p - \sigma_a)$ and τ_b depend on the soil type and strength, such that:

• for undrained conditions — fine grained clay soils, mixed soils and unweathered intact rock:

$$\left(\sigma_p - \sigma_a\right) = 2s_u, \text{ and}$$

$$\tau_b = s_u$$

where s_u = undrained shear strength.

- for drained conditions — uncemented sands and gravels and highly fractured rock:

$$\left(\sigma_p - \sigma_a\right) = \left(\gamma - \gamma_w\right) Z_c \left(K_p - K_a\right)$$

$$\tau_b = \left(\gamma - \gamma_w\right) Z_b \tan \phi'$$

where ϕ' = angle of friction,

$$K_p = \tan^2 \left(45° + \frac{\phi'}{2}\right) \text{ and}$$

$$K_a = \tan^2 \left(45° - \frac{\phi'}{2}\right)$$

- for drained conditions — cemented sands and gravels:

$$\left(\sigma_p - \sigma_a\right) = \left[\left(\gamma - \gamma_w\right) Z_c \left(K_p - K_a\right) + c' \left(K_{pc} + K_{ac}\right) \right]$$

$$\tau_b = c' + \left(\gamma - \gamma_w\right) Z_b \tan\phi'$$

where $K_{pc} = 2 \sqrt{K_p}$ and $K_{ac} = 2 \sqrt{K_a}$.

$(K_p - K_a)$ and $(K_{pc} + K_{ac})$ depend on ϕ' and values are given in Figure 10

Resistance to sliding is developed as shear stresses between the base of the thrust block and the soil, (Figure 11(a)). To fully mobilise this shearing resistance it is necessary to ensure rough contact between the base of the thrust block and ground. Alternatively, full base shear resistance can be mobilised by constructing a key (Figure 11(b)).

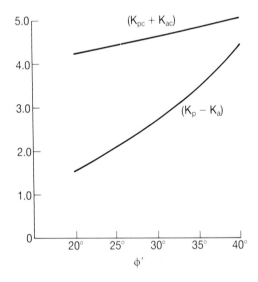

Figure 10 *Variation of earth pressure coefficients with friction angle*

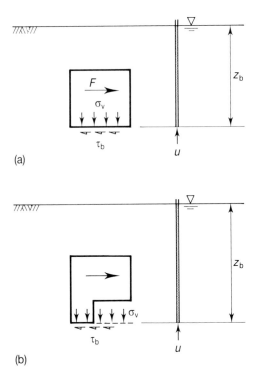

(a)

(b)

Figure 11 *Resistance to sliding*

NOTE:

1. For soils above the water table $\gamma_w = 0$; for soils below the water table $\gamma_w \approx 10$ kN/m^3.

2. When selecting values for Z_c and Z_b consider the possibility that part of the soil cover may be removed by natural erosion or by subsequent excavation or engineering works.

3. The active and passive earth pressure formulae are for smooth-faced blocks assuming no shear stress is mobilised between the block faces and the ground. *Alternative formulae to include shear stress should only be used with caution.*

5. The methods and formulae for calculating soil resistance apply only for cases where the lines of action of the applied forces and the ground resistance are approximately coincident, giving uniform stresses with no major moments or rotations.

2.9 STEP 8 – CALCULATE THRUST BLOCK SIZE TO RESIST HORIZONTAL FORCES

The size of the thrust block to resist horizontal forces must ensure that the design load is less than the ultimate resistance reduced by the appropriate reduction factor, i.e.

$$F_d < \frac{R_u}{T_r}.$$

2.10 STEP 9 – ESTIMATE THRUST BLOCK SIZE FOR UPTURN BENDS AND CALCULATE ULTIMATE BEARING RESISTANCE Q_b

The resultant thrust force for an upturn bend is equal to the sum of the vertical component of the fluid force, F_v, the weight of the block and soil cover. Since the weight of the thrust block is approximately the same as the weight of the same volume of soil the bearing capacity of the soil, q_b, is required to resist only the additional forces due to the fluid pressure. (see Section A4.5).

The ultimate bearing resistance, Q_b, can be calculated from the ultimate bearing capacity of the ground, q_b, below the thrust block.

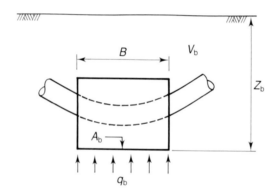

Figure 12 *Elements for calculating ultimate bearing resistance Q_b*

From Figure 12:

$$Q_b = q_b\, A_b$$

where A_b = base area of the thrust block, and

• for fined grained clay soils, mixed soils and unweathered intact rock:

$$q_b = 6s_u$$

where s_u is the undrained shear strength;

- for uncemented sands and gravels and highly fractured rock:

$$q_b = \frac{1}{2} \left(\gamma - \gamma_w \right) B N_\gamma + \left(\gamma - \gamma_w \right) \left(N_q - 1 \right) Z_b$$

where B is the minimum width of the thrust block;

- for cemented sands and gravels

$$q_b = c' N_c + \frac{1}{2} \left(\gamma - \gamma_w \right) B N_\gamma + \left(\gamma - \gamma_w \right) \left(N_q - 1 \right) Z_b.$$

N_c, N_γ and N_q are all related to the friction angle ϕ' as in Table 6:

Table 6 *Bearing capacity factors*

ϕ	N_c	N_γ	N_q
20°	14	3	6
25°	20	8	10
30°	30	18	18
35°	46	41	33

NOTE:

For soils above the water table $\gamma_w = 0$ and for soils below the water table $\gamma_w \approx 10$ kN/m^3.

2.11 STEP 10 – CALCULATE THRUST BLOCK SIZE TO RESIST DOWNWARD FORCES AT UPTURN BENDS

For upturn bends the vertical downward force must be less than the ultimate bearing resistance reduced by the appropriate reduction factor, i.e.

$$F_v < \frac{Q_b}{T_r}.$$

2.12 STEP 11 – ESTIMATE THRUST BLOCK SIZE FOR DOWNTURN BENDS AND CALCULATE THE EFFECTIVE WEIGHT W

The effective weight of the thrust block is either equal to:

$$W = V_b \left(\gamma_c - \gamma_w \right)$$

where γ_c is the unit weight of concrete and V_b the volume of concrete in the block, or

$$W = V_b \left(\gamma_c - \gamma_w \right) + V_s \left(\gamma - \gamma_w \right)$$

where γ is the unit weight of soil and V_s is the volume of soil above the block if it is appropriate to include it.

NOTE:

1. The weight of soil above the thrust block can only be used where it is certain that the soil cover will be present whenever the pipeline is under pressure.

2. If the block is above the water table then $\gamma_w = 0$.

3. Normal stresses on the base are reduced by the upward thrust forces and therefore base shear stresses are often neglected.

For downturn bends the vertical (upward) component of the design load must be less than the effective weight of the thrust block reduced by an appropriate factor of safety, F_s, which takes account of uncertainties in the calculation of the fluid thrust and the weights such that:

$$F_v < \frac{W}{F_s}$$

NOTE: It is suggested that the factor of safety, F_s, be not less than 1.5.

2.13 STEP 12 – DETERMINE MINIMUM THRUST BLOCK SIZE TO SUIT PIPE AND EXCAVATION DIMENSIONS

There is a minimum block size related to the minimum width and depth of trench required for the pipe to be placed, joints made, backfill compacted and to ensure the minimum statutory depth of cover (see BS 8010[1]).

> NOTE:
>
> Minimum cover of concrete around a pipe is usually not less than 200 mm increasing to 500 mm for larger pipe diameters.

2.14 STEP 13 – REVIEW THE DESIGN OF EACH THRUST BLOCK AND REVISE AS NECESSARY

The design procedure in Steps 1 to 12 should, in most cases, produce an acceptable and buildable design. If, however, the design is unacceptable, various assumptions can be reconsidered; the most significant of these are:

1. The ground properties (Step 5): re-evaluation of the ground properties may involve additional in-situ investigations and laboratory testing.

2. The thrust reduction factor (Step 6): re-evaluation of the thrust reduction factor may involve reassessment of the ground conditions and the importance of the construction.

If this reassessment fails to produce a reasonable design more fundamental considerations need to be addressed. These may include the route and geometry of the pipeline, the applicability of jointed pipes, the design and test pressures. The design will become a Class 2 problem.

2.15 STEP 14 – PREPARE DIMENSIONED DESIGN

To avoid local stress concentrations and to prevent rotation the thrust block should be proportioned so that the line of action of the fluid thrust forces passes through the middle third of the bearing face or, for vertical forces, through the middle third of the base area.

The dimensions of the block will depend on its purpose (bend, tee, end-stop, etc.). Guidance on suitable shapes for thrust blocks for a variety of conditions is given in Section 3.

3 Construction details

3.1 TYPICAL SHAPES FOR THRUST BLOCKS

The final shape, proportions and size of a thrust block are matters for the judgement of the designer in the light of the forces to be resisted and the prevailing ground conditions on site. Plane horizontal and vertical surfaces will usually be involved, but can be stepped where appropriate. The following diagrams indicate a variety of shapes which have been found useful.

3.1.1 Horizontal bends

A typical plan and section are illustrated in Figures 13(a) and 13(b). In poor ground when large forces are to be restrained keys below the main block, with reinforcing, and the use of piles may be considered, as illustrated by Figures 13(c), 13(d) and 13(e).

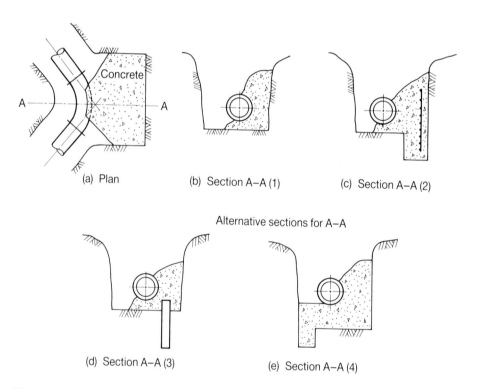

Figure 13 *Typical thrust block details for horizontal bends*

For very large pipes, those of 1.2 m diameter and above, the thrust blocks may consist of a reinforced concrete raft and a superstructure similar to that shown in Figure 14.

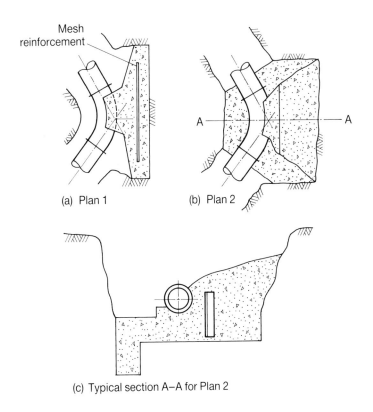

(a) Plan 1 (b) Plan 2

(c) Typical section A–A for Plan 2

Figure 14 *Typical details for thrust blocks for large pipes/thrust forces*

3.1.2 Vertical bends – upturns

The ideal bearing surface will usually be in a horizontal, or near horizontal plane. Keyed rafts may provide a solution when large diameter pipes and/or large forces and weak ground are encountered. Figures 15(a) to 15(d) illustrate some typical solutions.

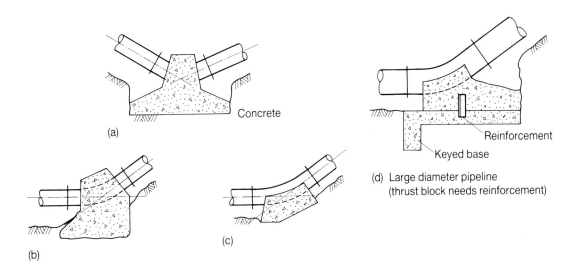

Figure 15 *Typical thrust block details for vertical bends (upturns) with downward thrust forces*

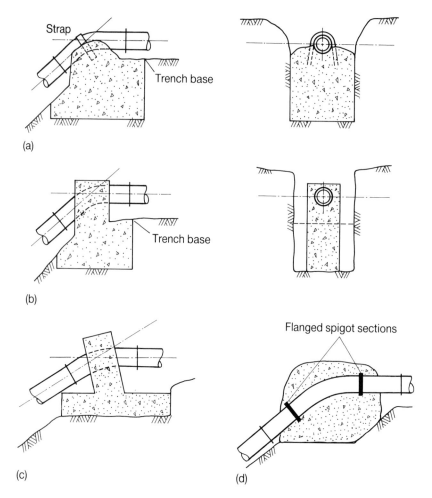

Figure 16 *Typical thrust block details for vertical bends (downturns) to resist uplift thrust forces*

3.1.3 Vertical bends – downturns

In this case the principal restraint is the weight of the block, plus the ground above it if appropriate. It is essential that the thrust block is actually attached to the pipeline, either by galvanised (or similarly protected) straps or by casing the block around the pipe. In this latter case it is important to maintain access to the joints. Figures 16(a) to 16(c) illustrate various typical sections for a thrust block to restrain a downturn vertical bend.

Occasionally the thrust block will be too large to allow easy clear access to the flexible joints on the pipe either side of the bend. Using flanged upright sections the bend section can be extended to ensure the flexible joints remain accessible. Figure 16(d).

3.1.4 Tee junctions

Equal tees give rise to fluid thrusts co-axial with the tee branch and hence the block should be mounted behind the tee-piece for horizontal connections or beneath it for vertical tees.

Unequal tees will also generate a force component co-axial with the main through flow, its magnitude being dependent on the area change. Figures 17(a) to 17(c) illustrate some typical sections for a thrust block designed to restrain a tee component.

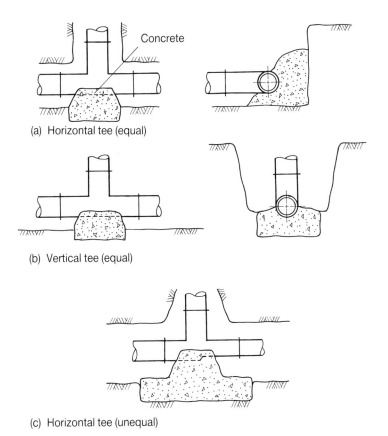

(a) Horizontal tee (equal)

(b) Vertical tee (equal)

(c) Horizontal tee (unequal)

Figure 17 *Typical thrust block details for tee-junctions*

Tees on very large pipelines, those of 1.2 m and above, may be supported on rafts. The raft would be laid first with the main structural supports cast into it. The final superstructure is cast at the time the pipeline is laid. A step or key, as illustrated in Figure 18, may be used to increase resistance to thrust forces.

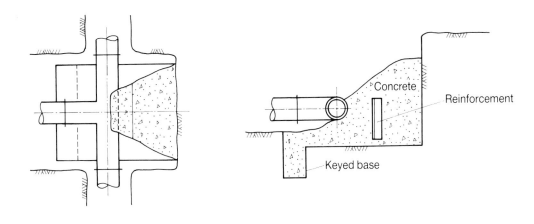

Figure 18 *A thrust block solution for a large diameter tee section or where thrust forces are very large*

3.1.5 Tapers

The hydraulic thrust to be restrained is co-axial with the pipe. Firm ground is needed, which will usually be found either at the side of the trench or at the bottom, against which the bearing surfaces of the thrust block can be cast.

The preferred solution is to cast the thrust block around the taper and into the firm ground either side of the taper. This avoids the creation of bending moments.

Additional restraint can be mobilised by keys below a block beneath the pipeline, but the superstructure must be sufficiently rigid to limit bending deflections. Figures 19(a) and 19(b) show typical sections and alternative plan arrangements for providing restraint for tapers.

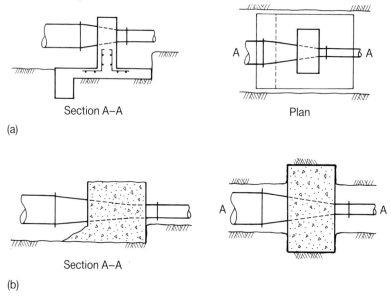

Figure 19 *Typical thrust block details for taper sections*

3.1.6 Sloping pipelines

Where buried pipes are laid in a straight line on slopes a component of the dead weight of the full pipeline acts axially, increasing with the angle of the slope. This axial force tends to encourage the pipes to slide down the slope, and the design must prevent such movement occurring.

On shallow slopes a buried pipe will usually be prevented from sliding by the frictional resistance of the backfill acting on the pipe wall, but the practice of wrapping the pipe loosely with polyethylene sleeving for additional protection against corrosion will significantly reduce the frictional resistance. Where slopes generate sliding forces greater than the available frictional resistance of the soil then the need for support structures will arise. These can vary from simple concrete thrust walls cast perpendicular to the pipe axis, to concrete drag anchors, possibly assisted by integral keys or even by raking piles, although such provisions will not usually be necessary on slopes of less than 1 in 4.

Thrust walls surrounding the pipe should extend at least half the pipe diameter above the crown and below the underside of the pipe and beyond the trench walls into undisturbed ground on either side, and be of suitable thickness to develop the required bond and to accept the shear and bending moments generated (see Figure 20).

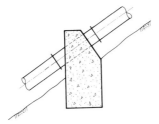

Figure 20 *Thrust block on a sloping pipeline*

NOTE: Any construction on a slope will alter the loading and groundwater conditions and may cause overall instability of the slope.

Pipes should be laid with their sockets facing uphill, and support structures located so that the external shoulder of the socket of each pipe bears against the pipe support. In this way puddle flanges or other securing devices are not required. It is important that each pipe is secured so that the tendency to slide does not cause movement which allows any joint to become fully 'homed'. Localised stresses can be high, caused by the relative settlement of the pipeline bedding material, the block and the cover to the pipe. These stresses can be critical for plastic pipes. The use of anchored or self-restrained joints as an alternative should be considered irrespective of the pipe materials.

If the sloping trench bottom is of a material that will form a drainage path for water, attention should be given to preventing the erosion of the bedding material beneath the pipe. The presence of cross thrust walls for support at each joint will tend to prevent large scale migration of the bedding support materials.

On long slopes, and depending on the gradient, more than one thrust block will be required. Table 7, taken from recommendations by Stanton Pipes[2] for cast iron pipelines, gives spacing for the thrust blocks.

Table 7 *Spacing for thrust blocks on long slopes*

Gradient	Spacing for thrust blocks
1 in 2	5.5 m
1 in 3	11.0 m
1 in 4	11.0 m
1 in 5	16.5 m
1 in 6	22.0 m

3.1.7 Blank ends

As with tapers, the hydraulic thrust is co-axial with the pipe but can seldom be directly restrained. Cutting into the side of the trench to firm ground and casting two blocks is usually the best solution. The end cap on the pipeline can then be held in place by a beam and chocks of hardwood (see Figure 21).

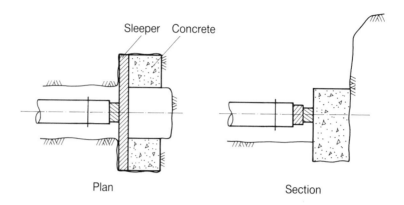

Sleeper   Concrete

Plan   Section

Figure 21 *Thrust block detail for a blank end or as a temporary measure for pressure testing*

3.2 INSTALLATION

The following section highlights some of the considerations and decisions required when installing jointed buried pressure pipelines in trenches with thrust blocks. Sound methods in design and attention to detail during construction are essential for reliability in service. Pipes should be laid on correctly prepared surfaces at the design grades with all fittings, valves, and similar components, installed as recommended by manufacturers.

3.2.1 Trenches

Trenches should be designed and excavated in accordance with good site practice, such as described in CIRIA Report 97 *Trenching Practice*[2].

The width of the trench at the surface is likely to vary according to its depth, type of soil, and method of excavation. At the pipe centre-line the trench width should allow access for the assembly of joints, connections of components such as valves, etc., and enable backfill material to be placed and compacted.

Dewatering, and the removal of running or standing water, is essential to avoid pipeline flotation and to enable suitable bearing surfaces to be prepared for the anchor blocks (see CIRIA Reports 97[3] and 113[4]).

Ideally the trench bottom should be firm but rough enough to maximise shear resistance, and the sides should be cut back to firm, undisturbed ground with a plane vertical surface normal to the direction of the applied load.

Sheet piling may be used to stabilise the side of the trench and ground against which the load bearing surfaces of the thrust blocks will be in contact. Adequate provision should be made to limit disturbance to the ground when the sheet piling is withdrawn. Normally, the piling behind thrust blocks should be withdrawn while the concrete is still plastic. Should later withdrawal be unavoidable and the ground be disturbed, then all backfilling should be done with mass concrete.

3.2.2 Concrete and reinforcement

The concrete for a thrust block should be dense and durable. A 25 Newton mix will normally suffice. Usually no (or merely nominal) reinforcement is needed except where tensile stresses may occur near anchorages for straps or to maintain the structural integrity of a very large block.

The ground should be tested for aggressive chemical conditions, particularly sulphates and special cements used as appropriate.

3.2.3 Thrust block – pipe bearing area

Good contact between the pipe and its supporting thrust block must be maintained and adequate provision made for differential expansion and contraction under load and from temperature changes.

For bends, tees, etc., especially in a horizontal plane, it is recommended that the bearing contact area should be limited to a quarter of the circumference: i.e. 45° either side of the pipe in the direction of the thrust through the centre of the pipe (see Figure 22).

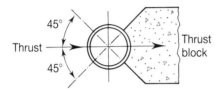

Figure 22 *Idealised bearing area for thrust block*

This is recommended particularly for unreinforced blocks to avoid cracking of the concrete from pipe deformation under load. Within the life of the pipeline pressures will fluctuate due to normal and transient events and it is recommended that the pipe be kept in close proximity to the block when unloaded by straps. Care should be taken to note any manufacturers' recommendations for supports, especially with regard to flexible pipes, in order to avoid creating local stress concentrations which might lead to fatigue failure of the pipe.

In those situations where it is desirable for the concrete to completely encircle the pipe wrapping or coating the pipe in a compressible material is highly recommended and indeed essential for thermoplastic pipes such as uPVC. For GRP pipes a compressible material is recommended at the end of the concrete only, approximately 50-75 mm into the concrete surround. If the wrapping is too flexible stress concentrations can be transferred to the internal junction between the concrete and the wrapping (see Figure 23).

The concrete anchor blocks should be shaped to leave the joint area clear for inspection and maintenance. Preferably, the pipe joint should protrude outside the flat face of the concrete.

If chambers, tank walls, and similar structures are being used to provide axial restraint, flanges can be cast, welded, or laminated onto the pipe wall. The external surface of the pipe wall may also be treated to enhance its bond with the concrete. If lateral forces are to be restrained and bending of the pipework external to the concrete is likely, flexible wrapping of the pipe, as described above is required for the region where the pipe enters the concrete to distribute stresses and avoid high local concentrations.

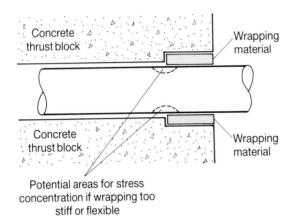

Figure 23 *Protection for fully enclosed uPVC pipes*

3.2.4 Shapes for improved capacity

If a key is constructed below a thrust block the effective size of the block is increased as shown in Figure 24, but the key should be placed on the side from which the force acts.

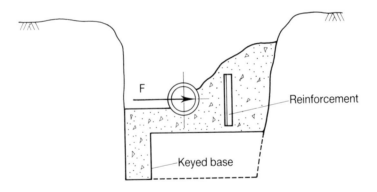

Figure 24 *Example of a keyed anchor block on a horizontal bend*

If the design of a thrust block is such that passive resistance of the soil is to be mobilised to restrain a non-axial force, then the face of the thrust block bearing against the soil should be perpendicular to the line of the force. In the case of a bend in a horizontal pipeline, where self-weight is negligible, the force will be horizontal and the face of the thrust block bearing against the soil should be vertical and perpendicular to the force.

Tapers and reducers experience net axial forces from the larger to the smaller diameter (see Figure 19). They need to be restrained around their whole circumference against axial movement. In some cases the thrust load will be transferred to the ground as a frictional load, possibly assisted by a key or piles below the block. Another method is to excavate into the side of the trench and transfer the load to undisturbed ground on both sides. The advantage of this method is that the net restraining force is collinear with the fluid force and hence the anchor block does not have to resist any moment.

3.2.5 Valves, valve chambers, pump houses and multiple duty thrust blocks

Partially closed valves will give rise to axial thrusts which need supporting similarly to tapers, except that under no-flow conditions the pressures either side of a partially closed valve will balance.

For a closed valve, the obstructed area is equal to the internal cross-sectional area of the valve gate plus the additional area associated with the joints and any upstream taper or step, if present.

A closed valve must be restrained to prevent it moving in the direction of thrust, usually by using a flanged valve with flanged spigot and puddle flange bearing onto a concrete thrust wall, normally the end wall of a concrete valve chamber. Valves installed without a chamber may be restrained, in much the same way, by using a concrete cross thrust wall.

In pump houses and concrete valve chambers the forces produced by opening and closing the valve(s) can be transmitted to the chamber walls and floor by mass concrete blocks or by means of flange-jointed pipework and puddle flanges (see Figure 25).

If a thrust block is used, then it will bear against the chamber walls and floor, which in turn transfer the forces to the ground. The size of the thrust block is determined using the same design procedures as for a thrust block bearing directly onto the ground. The chamber must be able to transfer the forces to the ground accommodating any bending moments and, if appropriate, submergence conditions.

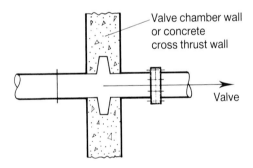

Figure 25 *Valve restraint by flanged spigot or puddle flange*

The chamber wall must be designed as a thrust wall, extending into undisturbed ground on either side of the chamber and, if necessary, below the level of the chamber floor. Reinforcement may be required to maintain structural integrity and provide resistance to bending moments. For an in-line isolating valve, the chamber will need to be constructed with a concrete wall. Only if the isolating valve is provided with an upstream flange adaptor may the remaining three walls be constructed in brickwork.

If the chamber floor is of concrete and integral with the end-thrust end wall, or thrust side wall in the case of a branch valve or tee, then resistance can be enhanced by keys beneath the floor perpendicular to the line of thrust. The force on the back of the tee, associated with a branch valve, in a chamber must be transferred by a thrust block to the side wall, which must be designed accordingly.

In a ring main, valves can be pressurised in either direction so the chamber walls must all be of concrete as both end walls need to resist movement in either direction. The chamber side walls are required to act as struts if the pipework contains flexible couplings to allow removal of a flanged valve.

Wings on chamber end walls may be used to provide the necessary bearing area and/or centralise the thrust. If over-excavation for a chamber occurs and the design utilises passive resistance on thrust walls, then backfilling with lean concrete must be specified. In poor ground, resistance may be increased by using keys or, in extreme cases, by raking piles.

A wash-out valve at a trough or an air valve at a peak can produce vertical forces which should be transferred into the ground through the chamber floor. Wash-out valves will also generate horizontal forces on the back of the branch tee when open and, if flexibly jointed to the tee, require restraining when shut. An air valve with its attendant isolating valve and tee will not usually generate any net horizontal force.

3.2.6 Embedded joints

Rubber ring joints should not be built into a thrust block. By definition the thrust block must be in place, and cured, before the test pressure is applied to prove the water tightness of the pipeline, and a badly made joint will leak inside the block if built-in.

3.2.7 Thrust block installed before the pipe

If the thrust block is to be poured before the pipeline is laid, then a concrete backfilling between the pipe wall and the face of the block should be included in the design. This will allow precise positioning of the bend or tee against the block during laying.

3.2.8 Thrust blocks on very large pipes

Thrust blocks at bends can become very large for large diameter pipes under high pressure. An effective restraining structure can be obtained by designing it as a reinforced raft beneath the pipe carrying a large key on the edge furthest from the thrust face. One or more vertical rolled steel joists cast into the raft and surrounded by concrete are used to restrain the thrust. The whole structure can be constructed before the pipe is laid, and the raft can be trapezoidal in plan for further economy.

3.2.9 Uplift due to groundwater

The presence of groundwater in the soil beneath the pipe may or may not cause insufficient ground support (or bearing capacity). When assessing the ground a final position of the groundwater table must be assumed to determine the relevant soil properties for design.

If the groundwater table is high and the pipeline is wholly or partially submerged, the trench should be dewatered to prevent the pipe becoming buoyant. Uplift from buoyancy forces on partially or wholly submerged concrete thrust blocks must be considered, especially where the forces being restrained have vertical components.

Groundwater levels are rarely constant and if they can rise above the invert level then the pipeline and associated thrust blocks must be designed for submerged conditions.

3.2.10 The 'Second Comer' (alterations and additions)

Situations can arise where a pipeline is installed in a trench alongside an existing pipe, or, in the case of a new pipe, provision may need to be made for alterations and for an adjacent pipeline.

For example, if a tee is installed on an existing pipeline then a suitable thrust block will be required as part of the new installation. Blanked off tees forming part of an initial installation for a later extension should have thrust blocks constructed behind them in anticipation of the expected test pressure in the branch when it is eventually laid.

Should a second pipeline be laid alongside a first in a common or adjacent trench after the first has been commissioned and is in operation, careful consideration must be given to the security of the first pipeline if it is to be in service while the ground is opened for the second time.

Twin lines through a bend are sometimes restrained by a common thrust block. Attention must be paid to whether the first part of the block, if cast in two parts, will be adequately restrained during excavation for the second part. Conversely, if the whole block is cast all at once, the designer must consider how the second pipeline will be laid through or against it.

There are circumstances when it is already known that there will be adjacent excavations for other services. When this is the case, the thrust block must not be cast against the ground but against a well founded substitute, usually sheet piling. The depth to which the sheet piling must be driven below the invert of the pipe depends on the forces, the ground conditions, and the depth of secondary excavation necessary for the installation of the adjacent services. To provide sufficient support, particularly if the invert depth of the adjacent service(s) is below the pipeline then initial excavations below the pipe inlet level may be necessary at the thrust block position to enable the sheet piling to provide the required support (see Figure 26).

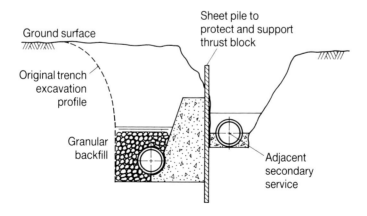

Figure 26 *Protecting the integrity of thrust blocks from adjacent excavations*

3.2.11 Poor ground

In ground of poor bearing capacity and where the passive resistance is low, the design of the thrust block may include vertical or raking piles. In this case the soil investigation should be carried to a depth greater than the depth of the piles. Alternatively, poor ground should be excavated to firm ground and backfilled with mass concrete or compacted granular fill.

When the force to be restrained acts axially, as at an end cap, the block may be built in the trench already excavated. Although no passive resistance mobilisation will be possible in unconsolidated trench backfill, restraint is provided by frictional resistance to sliding. If this is insufficient then the thrust block can take the form of a wall, or reinforced horizontal beam, extending into undisturbed ground on either side of the trench, symmetrical about the centre-line of the pipeline (see Figures 19(b) and 21).

In the absence of a chamber a thrust wall can be built to hold an in-line isolating valve against movement when shut. This thrust block, or wall, should be located on the opposite side of the valve to that which is pressurised. Restraint on both sides may be required if, under different operational conditions, either side can be pressurised and depressurised (see Figure 27).

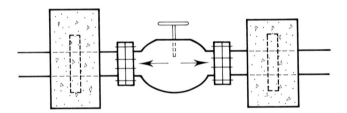

Figure 27 *Use of thrust walls in place of a chamber to restrain a valve*

The use of chambers to contain valves above 200 mm diameter should be considered. Servicing is easier and expensive assets are protected. The chamber walls can then provide the necessary restraint.

Thrust blocks for vertical forces at gradient changes will be designed to possess sufficient horizontal base bearing area to transfer the force into the ground in the case of a trough, and sufficient weight in the case of a peak. Since a peak may carry an air valve, and a trough a wash-out, the thrust blocks may also be chambers. At peaks, a gradient change will require the pipes to be physically strapped down to, or built into, the thrust block leaving the joints exposed.

4 Examples

4.1 INTRODUCTION

To show how the guide can be used, two demonstration examples are given. These consider the location of thrust blocks on a pipeline, the classification of the design, design calculations and specifications for a number of individual thrust blocks for a variety of components.

The final size, shape and construction details for the thrust blocks in these examples will depend on a number of factors, such as:

• ground conditions and ease of excavation

• restrictions on the use of, and type of, excavation plant

• availability of concrete on site as readymix.

Similar face and base area values can be achieved by a variety of block dimensions. Guidance on typical shapes and on the need to make joints available for inspection and maintenance, wherever possible, can be found in Section 3 of this guide.

4.2 EXAMPLE 1

Figure 28 shows a plan view of a pipeline of nominal diameter 250 mm and an outside diameter of 310 mm, laid to minimum gradients (i.e. roughly horizontal) between location 1 and 4. The branch line 3 to 5, having a nominal diameter of 150 mm and an outside diameter of 194 mm, is also approximately horizontal. The nominal working pressure is 3 bar and under surge conditions the maximum pressure is 3.8 bar. Ground conditions are typical of the glacial drift found in many midland and northern areas of the UK.

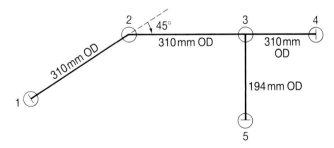

Figure 28 *Plan view of pipeline*

From a walkover survey of the pipeline route and hand samples from trial pits adjacent to the route the following characteristic values for strength and density were assessed using Table 3 as a guide:

Table 8 *Ground conditions for pipeline in example 1*

Location	Description	Type	s_u kPa	c' kPa	ϕ'	γ kN/m³
1	Firm to stiff clay	undrained	100	N/A	N/A	20
2	Firm clay	undrained	75	N/A	N/A	20
3	Soft clay and peat					
4	Medium dense sand					
		drained	N/A	0	35°	17
5	Lightly cemented sand	drained	N/A	10	35°	18

It was also noted that groundwater levels were well below the maximum depth of excavation and therefore $\gamma_w = 0$ in all design calculations.

Step 1 is to locate the positions where thrust blocks are required and 5 locations are identified on Figure 28. For this example the nominal working pressure is 3 bar and the maximum pressure, under surge conditions, is 3.8 bar and hence, following the recommendations of BS 8010[1], the test pressure is 3.8 + 5 = 8.8 bar.

Detailed design of each thrust block will be considered separately.

4.2.1 Design for the thrust block at location 1

Step 2. The design force F_d is obtained from Table 1 or Figure 6 or calculated directly using the appropriate formula in Figure 5.

At location 1 the component is a dead end and

$$F_d = pA = (8.8 \times 10^2) \frac{\pi}{4} \times 0.31^2 = 66.4 \text{ kN}$$

(Note $F_d < 1000$ kN)

Step 3. From the information given in Table 8 the soils at location 1 are described as firm to stiff clay and, from Table 2, the conditions are good ground.

(Note that the design is in Class 1).

Step 4. The direction of the design force is horizontal.

Step 5. From the information given in Table 8 the soil is undrained and the appropriate design parameters are $s_u = 100$ kPa and $\gamma = 20$ kN/m³.

Step 6. From Table 5, for firm to stiff clay take a value for the reduction factor $T_r = 3$.

Step 7. For a thrust block with a face area $A_f = 0.75$ m² and a base area $A_b = 0.50$ m² the ultimate resistance, from Section 2.7 is

$$R_u = 2s_u A_f + s_u A_b = 2 \times 100 \times 0.75 + 100 \times 0.5 = 200 \text{ kN}$$

Step 8. From the above $F_d = 66.4$ kN and $\dfrac{R_u}{T_r} = \dfrac{200}{3} = 66.7$ kN so the design is satisfactory

Step 12. The external diameter of the pipe is 310mm: if the minimum cover around the pipe is 300 mm the minimum area is

$$A_f = (2 \times 300 + 310)^2 \times 10^{-6} = 0.82 \text{ m}^2$$

Although this is a little larger than the area $A_f = 0.75$ m^2 in Step 7 the larger area would probably be specified to ensure buildability.

4.2.2 Design for the thrust block at location 2

Step 2. The design force F_d is obtained from Table 1 or Figure 6 or calculated directly using the appropriate formula in Figure 5.

At location 2 the component is a 45° bend and

$$F_d = 2pA \sin \frac{\theta}{2} = 2 \times (8.8 \times 10^2) \frac{\pi}{4} \times 0.31^2 \sin \left(\frac{45}{2}\right) = 50.8 \text{ kN}$$

(note $F_d < 1000$ kN).

Step 3. From the information given in Table 8 the soils at location 2 are described as firm clay and, from Table 2 the conditions are good ground.

(Note the design is Class 1).

Step 4. The direction of the design force is horizontal.

Step 5. From the information given in Table 8 the soil is undrained and the appropriate design parameters are $s_u = 75$ kPa and $\gamma = 20$ kN/m^3.

Step 6. From Table 5, for firm clay take a value for the thrust reduction factor $T_r = 4$.

Step 7. For a thrust block with a face area $A_f = 1$ m^2 and a base area $A_b = 0.75$ m^2 the ultimate resistance from section 2.7 is

$$R_u = 2s_u A_f + s_u A_b = 2 \times 75 \times 1 + 75 \times 0.75 = 206 \text{ kN}$$

Step 8. From the above $F_d = 50.8$ kN

and $\dfrac{R_u}{T_r} = \dfrac{206}{4} = 51.5$ kN so the design is satisfactory.

Step 12. As for location 1 (above) the minimum area A_f is 0.82 m^2 so the design $A_f = 1$ m^2 is a little larger than the minimum.

4.2.3 Design for the thrust block at location 3

Step 2. The design force F_d is obtained from Table 1 or Figure 6 or calculated directly using the appropriate formula in Figure 5.

At location 3 the component is a tee connection and

$$F_d = pA_2 = (8.8 \times 10^2) \frac{\pi}{4} \times 0.194^2 = 26.0 \text{ kN}$$

(Note $F_d < 1000$ kN).

Step 3. From the information given in Table 8 the soils at location 3 are described as soft clay and peat and, from Table 2, the conditions are poor ground.

The design is Class 2 (because of the poor ground conditions). Either, geotechnical and hydraulic experts should be consulted, or the poor ground excavated and replaced with engineered fill or the pipeline and component relocated to avoid the area of poor ground.

4.2.4 Design for the thrust block at location 4

Step 2. The design force F_d is obtained from Table 1 or Figure 6 or calculated directly using the appropriate formulae in Figure 5.

At location 4 the component is a dead end and

$$F_d = pA = (8.8 \times 10^2) \frac{\pi}{4} \times 0.31^2 = 66.4 \text{ kN}$$

(Note $F_d < 1000$ kN).

Step 3. From the information given in Table 8 the soils at location 4 are described as medium dense sand above the water table and, from Table 2, the conditions are good ground.

(Note that the design is in Class 1).

Step 4. The direction of the design force is horizontal.

Step 5. From the information given in Table 8 the soil is drained and the appropriate design parameters are $\phi' = 35°$ and $\gamma = 17$ kN/m^3.

Step 6. From Table 5, for medium dense sand, take a value for the reduction factor $T_r = 4$.

Step 7. For a thrust block with a face area $A_f = 2.5$ m^2 and $A_b = 1$ m^2 the ultimate resistance from Section 2.7 is

$$R_u = (\gamma - \gamma_w) Z_c (K_p - K_a) A_f + (\gamma - \gamma_w) Z_b \tan \phi' A_b$$

For soil above the water table $\gamma_w = 0$; for $\phi' = 35°$, from Figure 10, $(K_p - K_a) = 3.3$. If the depth of the axis of the pipe is 2 m then $Z_c = 2$ m and $Z_b = 2.5$ m.

$$R_u = 17 \times 2 \times 3.3 \times 2.5 + 17 \times 2.5 \times \tan 35° \times 1 = 310 \text{ kN}$$

Step 8. From the above $F_d = 66.4$ kN and $\dfrac{R_u}{T_r} = \dfrac{310}{4} = 78$ kN so the design is satisfactory

Step 12. As for location 1 (above) the minimum area A_f is 0.82 m^2 and so the design $A_f = 2.5$ m^2 is significantly larger than the minimum.

4.2.5 Design for the thrust block at location 5

Step 2. The design force F_d is obtained from Table 1 or Figure 6 or calculated directly using the appropriate formula in Figure 5. The design pressure and pipe area are the same as those at location 3 and

$$F_d = 26.0 \text{ kN}$$

(Note $F_d < 1000$ kN)

Step 3. From the information given in Table 8 the soils at location 5 are described as lightly cemented sand and, from Table 2, the conditions are good ground.

(Note the design is in Class 1).

Step 4. The direction of the design force is horizontal.

Step 5. From the information given in Table 8 the soil is drained and the appropriate design parameters are $c' = 10$ kPa, $\phi' = 35°$ and $\gamma = 18$ kN/m^3.

Step 6. From Table 5, for lightly cemented soil, take a value for the reduction factor $T_r = 4$.

Step 7. For a thrust block with a face area $A_f = 0.5$ m^2 and $A_b = 0.5$ m^2 the ultimate resistance from Section 2.7 is

$$R_u = \left[(\gamma - \gamma_w) \, Z_c \, (K_p - K_a) + c'(K_{pc} + K_{ac}) \right] A_f + \left[c' + (\gamma - \gamma_w) \, Z_b \, \tan\phi' \right] A_b$$

For soil above the water table $\gamma_w = 0$; for $\phi' = 35°$; from Figure 10, $(K_p - K_a) = 3.3$ and $(K_{pc} + K_{ac}) = 4.8$. If the depth of the axis of the pipe is 2 m then $Z_c = 2$ m and $Z_b = 2.5$ m.

$$R_u = [18 \times 2 \times 3.3 + 10 \times 4.8] \, 0.5 + [10 + 18 \times 2.5 \tan 35°] \, 0.5 = 104 \text{ kN}$$

Step 8. From the above $F_d = 26.0$ kN and

$$\frac{R_u}{T_r} = \frac{104}{4} = 26.0 \text{ kN} \quad \text{so the design is satisfactory.}$$

Step 12. The external diameter of the pipe is 194 mm; if the minimum cover around the pipe is 300 mm the minimum area is $A_f = (2 \times 300 + 194)^2 \times 10^{-6} = 0.63$ m^2

The design $A_f = 0.5$ m^2 is a little smaller than the minimum.

Step 13. With the exception of location 3, where the ground conditions are poor and the situation is in Class 2, the design sizes appear reasonable. For location 3 the various alteratives discussed above should be considered.

4.3 EXAMPLE 2

Figure 29 shows sections of a pipeline in undulating terrain. The main pipeline 1-5 is straight in plan. The outside diameter of the main pipeline 1-5 is 700 mm and the outside diameter of the branch is 450 mm. The elevations are the invert of the pipe. The ground conditions are known to be stiff clay throughout (take $s_u = 100$ kPa and $\gamma = 20$ kN/m^3).

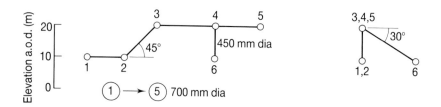

Figure 29 *Pipeline for Example 2*

Step 1 is to locate the positions where thrust blocks are required and these are given in Table 9. The test pressure will be 25 bar and this will be applied at location 5 so that the pressures at locations 3, 4 and 5 will be 25 bar and the pressures at locations 1, 2 and 6, which are 10 m lower, will be 26 bar.

Table 9 *Location and type of thrust blocks (see Figure 29)*

Location	Type
1	Blank end
2	Upturn bend
3	Downturn bend
4	Tee
5	Blank end
6	Blank end

(Note that at locations 2, 3, 4 and 6 the thrusts have both horizontal and vertical components)

4.3.1 Design for the thrust block at location 1

Step 2. The design force F_d is obtained from Table 1 or Figure 6 or calculated directly using the appropriate formulae in Figure 5, taking a test pressure of 26 bar and a diameter of 0.70 m.

At location 4 the component is a blank end and

$$F_d = pA = (26 \times 10^2)\ \frac{\pi}{4} \times 0.70^2 = 1001\ \text{kN}$$

(Note $F_d \approx 1000$ kN)

Step 3. The ground is stiff clay and so the conditions are good ground

(Note that the design can be taken to be in Class 1)

Step 4. The direction of the design force is horizontal.

Step 5. From the information given above the soil is undrained and the appropriate design parameters are $s_u = 100$ kPa and $\gamma = 20$ kN/m^2.

Step 6. From Table 5, for stiff clay takes a value for the reduction factor $T_r = 2.5$. (This is the mid-point of the range 2 to 3 given in Table 5).

Step 7. For a thrust block with sides 3 m the face area is $A_f = 9$ m^2 and the base area is $A_b = 9$ m^2. From section 2.7 the ultimate resistance is

$$R_u = 2s_u A_f + s_u A_b = 2 \times 100 \times 9 + 100 \times 9 = 2700 \text{ kN}$$

Step 8. From the above $F_d = 1001$ kN

and $\dfrac{R_u}{T_r} = \dfrac{2700}{2.5} = 1080$ kN so the design is satisfactory.

Step 12. The external diameter of the pipe is 700 mm; if the minimum cover around the pipe is (say) 500 mm the minimum face area is

$$A_f = (2 \times 500 + 700)^2 \times 10^{-6} = 2.9 \text{ m}^2$$

This is considerably smaller than the design size $A_f = 9$ m^2 in Step 7.

4.3.2 Design for the thrust block at location 2

Step 2. The design force is obtained from Table 1 or Figure 6 or calculated directly from the appropriate formulae in Figure 5 taking a test pressure of 26 bar and a diameter 0.70 m.

At location 2 the component is a 45° upturn bend and there are horizontal and downward vertical components (see Figure 5).

The vertical component is

$$F_{dv} = pA \sin\theta$$
$$= (26 \times 10^2) \frac{\pi}{4} \times 0.70^2 \times \sin 45°$$
$$= 707 \text{ kN}$$

The horizontal component is

$$F_{dh} = pA (1-\cos\theta)$$
$$= (26 \times 10^2) \frac{\pi}{4} \times 0.70^2 \times (1-\cos 45°)$$
$$= 293 \text{ kN}$$

(The total resultant design force $F_d = \sqrt{(F_{dv}^2 + F_{dh}^2)} = 765$ kN)

Step 3. The ground is stiff clay and so the conditions are good ground.

(Note the design is in Class 1)

Step 4. The thrust block must resist both horizontal and vertical (downward) forces.

Step 5. From the information given above the soil is undrained and the appropriate design parameters are s_u = 100 kPa and γ = 20 kN/m³.

Step 6. From Table 5, for stiff clay take a value of the thrust reduction factor T_r = 2.5. (This is the mid-point of the range 2 to 3 given in Table 5).

Step 7. For a thrust block with a base 2 m × 1.5 m and 2 m high A_f = 4 m² and A_b = 3 m². From Section 2.7 the ultimate resistance for horizontal loading (neglecting the effect of base shear) is

$R_{uh} = 2s_u \, A_f = 2 \times 100 \times 4 = 800$ kN

Step 8. From the above F_{dh} = 293 kN

and $\dfrac{R_u}{T_r} = \dfrac{800}{2.5}$ = 320 kN so the design is satisfactory.

Step 9. From Section 2.9 the ultimate bearing capacity for vertical loading is

$Q_b = 6s_u \, A_b = 6 \times 100 \times 3 = 1800$ kN

Step 10. From the above F_{dv} = 707 kN and $\dfrac{Q_b}{T_r} = \dfrac{1800}{2.5}$ = 720 kN so the design is

satisfactory.

Step 12. As before the minimum face area for a 700 mm diameter pipe with 500 mm concrete cover is 2.9 m² which is smaller than the design size. A_f = 4 m² in Step 7.

4.3.3 Design for the thrust block at location 3

Step 2. The design force is obtained from Table 1 or Figure 6 or calculated directly from the appropriate formulae in Figure 5 taking a test pressure of 26 bar and a diameter 0.70 m.

At location 3 the component is a 45° downturn bend and there are horizontal and upward vertical components (see Figure 5).

The vertical component is

$$F_{dv} = pA \sin\theta$$
$$= (25 \times 10^2) \frac{\pi}{4} \times 0.70^2 \times \sin 45°$$
$$= 680 \text{ kN}$$

The horizontal component is

$$F_{dh} = pA (1-\cos\theta)$$
$$= (25 \times 10^2) \frac{\pi}{4} \times 0.70^2 \times (1-\cos 45°)$$
$$= 282 \text{ kN}$$

(The total resultant design force $F_d = \sqrt{\left(F_{dv}^2 + F_{dh}^2\right)} = 736 \text{ kN}$)

Step 3. The ground is stiff clay and so the conditions are good ground.

(Note the design is in Class 1)

Step 4. The thrust block must resist both horizontal and vertical (upward) forces.

Step 5. From the information given above the soil is undrained and the appropriate design parameters are $s_u = 100$ kPa and $\gamma = 20$ kN/m³.

Step 6. From Table 5, for stiff clay take a value for the thrust reduction factor $T_r = 2.5$. (This is the mid-point of the range 2 to 3 given in Table 5).

Step 7. For a thrust block with a base 5 m $\times$ 5 m and 4.3 m high $A_f = 21.5$ m² and the volume is $V_b = 107.5$ m³. From section 2.7 the ultimate resistance for horizontal loading (neglecting the effect of base shear) is

$$R_{uh} = 2s_u A_f = 2 \times 100 \times 21.5 = 4300 \text{ kN}$$

Step 8. From the above $F_{dh} = 282$ kN

and $\dfrac{R_u}{T_r} = \dfrac{4300}{2.5} = 1720$ kN so the design is satisfactory.

Step 11. From Section 2.11 (neglecting the weight of any backfill above the thrust block and taking the water table above the block) the effective weight is

$$W = V_b (\gamma_c - \gamma_w) = 107.5 \times 10 = 1075 \text{ kN}$$

Taking a value for the factor of safety $F_s = 1.5$ then, from the above $F_{dv} = 680$ kN

and $\dfrac{W}{F_s} = \dfrac{1075}{1.5} = 716$ kN and the design is satisfactory.

Step 12. As before the minimum face area for a 700 mm dia pipe with 500 mm concrete cover is 2.9 m^2 which is smaller than the design size.

4.3.4 Design for the thrust block at location 4

Step 2. The design force is obtained from Table 1 or Figure 6 or calculated directly from the appropriate formulae in Figure 5 taking a test pressure of 25 bar and a diameter 450 mm.

At location 4 the component is a tee junction and the design force F_d acts in the line of

the branch 6-4 and $\quad F_{dv} = pA = (25 \times 10^2)\ \dfrac{\pi}{4} \times 0.45^2 = 398$ kN

Resolving vertically and horizontally and components of the design force are

$F_{dv} = F_d \sin 30° = 200$ kN (acting upwards)
$F_{dh} = F_d \cos 30° = 347$ kN

Step 3. The ground is stiff clay and so the conditions are good ground.

(Note the design is in Class 1)

Step 4. The thrust block must resist both horizontal and vertical (upward) forces in the plane of the branch. There are no resultant forces in the line of the main 700 mm pipeline.

Step 5. From the information given above the soil is undrained and the appropriate design parameters are $s_u = 100$ kPa and $\gamma = 20$ kN/m^3.

Step 6. From Table 5, for stiff clay take a value for the thrust reduction factor $T_r = 2.5$. (This is the mid-point of the range 2 to 3 given in Table 5).

Step 7: For a thrust block with a face 3 m $\times$ 3 m and 3.3 m long $A_f = 9$ m^2 and the volume is 30 m^3. From section 2.7 the ultimate resistance for horizontal loading (neglecting base shear) is

$R_{uh} = 2s_u A_f = 2 \times 100 \times 9 = 1800$ kN

Step 8. From the above $F_{dh} = 347$ kN

and $\dfrac{R_u}{T_r} = \dfrac{1800}{2.5} = 720$ kN so the design is satisfactory

Step 11. From Section 2.11 (neglecting the weight of any backfill above the thrust block and taking the water table above the block) the effective weight is

$W = V_b (\gamma_c - \gamma_w) = 30 \times 10 = 300$ kN

Taking a value for the factor of safety $F_s = 1.5$ then, from the above $F_{dv} = 200$ kN

and $\dfrac{W}{F_s} = \dfrac{300}{1.5} = 200$ kN and the design is satisfactory.

Step 12. The external diameter of the pipe is 450 mm; if the minimum cover around the pipe is (say) 400 mm the minimum face area is

$A_f = (2 \times 400 + 450)^2 \times 10^{-6} = 1.56$ m^2

This is considerably smaller than the design size $A_f = 9$ m^2 in Step 7.

4.3.5 Design for the thrust block at location 5

The component, the loads and the ground conditions at location 5 are the same as those at location 1 except that the test pressure is 25 bar instead of 26 bar. This is a very small difference and the thrust block at location 5 can be the same size as the thrust block at location 1.

4.3.6 Design for thrust block at location 6

Step 2. The design force is obtained from Table 1 or Figure 6 or calculated directly from the appropriate formulae in Figure 5 taking a test pressure of 25 bar and a diameter of 0.45m.

At location 6 the component is a dead end and the design force F_d acts in the direction

of the branch 4-6 and $F_d = pA = (26 \times 10^2) \dfrac{\pi}{4} \times 0.45^2 = 413$ kN

Resolving vertically and horizontally and components of the design force are

$F_{dv} = F_d \sin 30° = 207$ kN (acting downwards)
$F_{dh} = F_d \cos 30° = 358$ kN

Step 3. The ground is stiff clay and so the conditions are good ground.

(Note the design is in Class 1)

Step 4. The thrust block must resist both horizontal and vertical (downward) forces.

Step 5. From the information given above the soil is undrained and the appropriate design parameters are $s_u = 100$ kPa and $\gamma = 20$ kN/m^3.

Step 6. From Table 5, for stiff clay take a value for the thrust reduction factor $T_r = 2.5$. (This is the mid-point of the range 2 to 3 given in Table 5).

Step 7. For a thrust block with a face 2.2 m × 2.2 m and 0.5 m long $A_f = 4.84$ m^2 and $A_b = 1.1$ m^2. From Section 2.7 the ultimate resistance for horizontal loading (neglecting the effect of base shear) is

$$R_{uh} = 2s_u A_f = 2 \times 100 \times 4.84 = 968 \text{ kN}$$

Step 8: From the above $F_{dh} = 358$ kN

and $\dfrac{R_u}{T_r} = \dfrac{968}{2.5} = 387$ kN so the design is satisfactory.

Step 9. From section 2.9 the ultimate bearing capacity for vertical loading is

$$Q_b = 6s_u A_b = 6 \times 100 \times 1.1 = 660 \text{ kN}$$

Step 10. From the above $F_{dv} = 207$ kN

and $\dfrac{Q_b}{T_r} = \dfrac{660}{2.5} = 264$ kN so the design is satisfactory.

Step 12. As before the minimum face area for a 450 mm diameter pipe with 400 mm concrete cover is 1.56 m^2 which is considerably smaller than the design size $A_f = 4.84$ m^2 in Step 7.

Step 13. All the thrust blocks are relatively large but so too are the forces to be resisted (up to 1000 kN or 100 tonnes). This arose from the relatively large pipe diameters and test pressure.

The largest blocks occur at locations 3 and 4 where there are upward forces to be resisted by the weight of the block. In these cases particularly more efficient and economic designs may be obtained using tension piles.

Where relatively large horizontal forces occur, particularly at locations 1 and 5, consideration could be given to use of short shear piles.

Designs using piles will require input from a specialist geotechnical engineer.

References

1. BRITISH STANDARDS INSTITUTION
   Code of practice for pipelines
   BS 8010

2. *Ductile Iron Pipelines – design information and data*
   Stanton PLC, Nottingham

3. IRVINE, D.J. and SMITH, R.J.H.
   Trenching practice
   CIRIA, Report 97, 1983

4. SOMERVILLE, S.H.
   Control of groundwater for temporary works.
   CIRIA Report 113, 1986

5. BRITISH STANDARDS INSTITUTION.
   Code of practice for site investigations.
   BS 5930 : 1981

Bibliography

Fluid Dynamics and Hydrostatics

STREETER, V.L. and WYLIE, E.B.
Fluid mechanics
McGraw Hill London 1988.

WYLIE, E.B. and STREETER, V.L.
Fluid transients in systems
Prentice Hall, Englewood Cliffs, N.J. 1993.

CHAUDHRY, M.H.
Applied hydraulic transients
Second Edition, Van Nostrand Reinhold 1987.

THORLEY, A.R.D.
Fluid transients in pipeline systems
D & L George Ltd
September 1991

THORLEY, A.R.D. and ENEVER, R.J.
Control and suppression of pressure surges in pipelines and tunnels
CIRIA Report 84, 1979 (Reprinted 1985).

Soil Mechanics and Ground Engineering

BRITISH STANDARDS INSTITUTION.
Methods of test for soils for civil engineering purposes
BS 1377 : 1975.

BRITISH STANDARDS INSTITUTION.
Code of practice for foundations.
BS 8004:1986

WELTMAN, A.J. and HEAD, J.M.
Site investigation manual.
CIRIA Special Publication 25/PSA Civil Engineering Technical Guide 35, 1983.

PADFIELD, C.J. and MAIR, R.J.
Design of retaining walls embedded in stiff clays.
CIRIA Report 104, 1984.

PADFIELD, C.J. and SHARROCK, M.J.
Settlement of structures on clay soils.
CIRIA Special Publication 27, 1983.

Appendix 1 Determination of fluid pressure forces

In the following sections the detailed derivation of the fluid forces and the required reactions from the supporting structures are developed for various components. A one-dimensional approach is used since this will be valid for virtually all cases of interest, e.g. even for a 2000 mm dia. pipeline the pressure variation over the vertical diameter will be only 0.2 bar.

A1.1 Steady fluid forces on bends in a horizontal plane

Pipe cross-sectional area $= A_1, A_2$

flow velocities $= v_1, v_2$

R_x, R_y are the components of the restraining force required

From conservation of linear momentum

$$\sum F_x = \rho Q \left(v_{2x} - v_{1x}\right) + \rightarrow :$$

$$p_1 A_1 - p_2 A_2 \cos\theta - R_x = \rho Q \left(v_2 \cos\theta - v_1\right) \tag{1}$$

From conservation of mass flow,

$$\dot{m} = \rho_1 A_1 v_1 = \rho_2 A_2 v_2 \tag{2}$$

If the pipe bend is of uniform diameter, $A_1 = A_2$ and for a liquid, density $\rho_1 = \rho_2$, equation [2] reduces to:

$$v_1 = v_2 = \frac{Q}{A} \tag{3}$$

From conservation of energy

$$\frac{1}{\rho g}\left(p_2 - p_1\right) + \frac{1}{2g}\left(v^2_2 - v_1^2\right) + (z_2 - z_1) + h_L = 0 \tag{4}$$

For most practical pipe bends the friction losses h_L may be regarded as small, to the point of being negligible, in relation to the pressure heads.

Similarly, for flow in a horizontal plane, the elevation change $(z_2 - z_1) = 0$ and from equation [3], $v_1 = v_2$. $\tag{5}$

Hence, it follows that $p_1 = p_2$ and substituting [3] and [5] in [1] yields:

$$R_x = pA (1 - \cos\theta) - \frac{\rho Q^2}{A} (\cos\theta - 1) \qquad [6]$$

Considering force components in the y-direction, from linear momentum considerations, as before

$$\sum F_y = \rho Q (v_{2y} - v_{1y}) \uparrow +:$$

$$R_y - p_2 A_2 \sin\theta = \rho Q (v_2 \sin\theta) \qquad [7]$$

Assuming uniform cross-sectional area, fluid density, and negligible losses as above, equation [6] reduces to:

$$R_y = \left(pA + \frac{\rho Q^2}{A} \right) \sin\theta \qquad [8]$$

Most water pipelines operate at relatively modest flow velocities and the dynamic head in the above equations is really quite small. For example the dynamic pressure head for water flowing at 3 m/s is less than 0.05 bar. Hence the final simplifications are:

From [6]: $\qquad R_x = pA (1 - \cos\theta) \qquad [9]$

From [8]: $\qquad R_y = pA \sin\theta \qquad [10]$

and the resultant thrust

$$R = \sqrt{R_x^2 + R_y^2} = 2pA \sin\frac{\theta}{2} \qquad [11]$$

A1.2 Steady fluid forces on a bend in a vertical plane

(a) Upturn

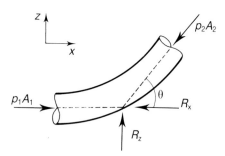

The reaction R_x in the horizontal direction will be modified due to the change in elevation affecting the pressures, i.e. $p_2 = p_1 - \rho g (z_2 - z_1)$ leading to:

$$R_x = p_1 A (1 - \cos\theta) + \rho g A (z_2 - z_1) + \frac{\rho Q^2}{A} (1 - \cos\theta) \qquad [12]$$

For small and medium sized bends the first term will be dominant,
i.e. $R_x = pA (1 - \cos\theta)$

In the vertical direction:

$$R_z - p_2 A_2 \sin\theta = \rho Q \left(v_2 \sin\theta\right) \qquad [13]$$

and since $v_2 = \dfrac{Q}{A_2} = \dfrac{Q}{A}$ for a uniform pipe

$$R_z = p_2 A \sin\theta + \frac{\rho Q^2}{A} \sin\theta \qquad [14]$$

The effect of the dynamic pressure head can again be assumed negligible for the cases of interest in this report. The working form of equation [14] is therefore:

$$R_z = pA \sin\theta \qquad [15]$$

for a bend having the configuration shown, i.e. an upturn.

(b) Downturn

$$R_z = pA \sin\theta \qquad [16]$$

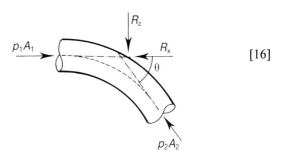

A1.3 Steady forces on a tee-junction

To set up the general equation an unequal tee is considered in a vertical plane.

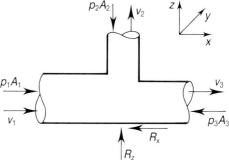

For conservation of linear momentum in the x-direction:

$$\sum F_x = \iint \rho \, v_x \, v \, . \, dA \quad +\rightarrow:$$

$$p_1 A_1 - p_3 A_3 - R_x = \rho \left[A_3 v_3^2 - A_1 v_1^2\right]$$

$$= \rho \left[\frac{Q_3^2}{A_3} - \frac{Q_1^2}{A_1}\right] \qquad [17]$$

therefore:

$$R_x = p_1 A_1 - p_3 A_3 - \rho \left[\frac{Q_3^2}{A_3} - \frac{Q_1^2}{A_1} \right] = p (A_1 - A_3) \qquad [18]$$

neglecting the trivial dynamic pressure and friction effects

Similarly in the z-direction, i.e. vertical:

$$\sum F_x = \iint \rho \, v_z \, v \, . \, dA \rightarrow +:= \frac{\rho Q_2^2}{A_2}$$

therefore,

$$R_z - p_2 A_2 = \rho . A_2 v_2^2 \qquad [19]$$

$$R_z = p_2 A_2 + \frac{\rho Q_2^2}{A_2} = p A_2 \qquad [20]$$

neglecting friction losses and the dynamic pressure.

A1.4 Forces on changes in section (tapers)

From conservation of linear momentum in the x-direction:

$$F_x = \rho Q \left(v_{2x} - v_{1x} \right) + \rightarrow :$$

$$p_1 A_1 - p_2 A_2 - R_x = \rho Q \left(v_2 - v_1 \right) \qquad [21]$$

giving $R_x = p_1 A_1 - p_2 A_2 - \rho A_1 v_1 \left(v_2 - v_1 \right) \qquad [22]$

From conservation of mass flow $\quad A_1 v_1 = A_2 v_2 \qquad [23]$

hence, $\quad R_x = p_1 A_1 - p_2 A_2 - \rho A_1 v_1^2 \left(\frac{A_1}{A_2} - 1 \right) \qquad [24]$

The two pressures can be related by examining the conservation of energy equation:

$$\frac{1}{\rho g} (p_2 - p_1) + \frac{1}{2g} \left(v_2^2 - v_1^2 \right) + h_L = 0$$

and, by neglecting viscous losses, i.e. $h_L = 0$

$$p_2 = p_1 - \frac{\rho}{2} \left(v_2^2 - v_1^2 \right)$$

$$= p_1 - \frac{\rho v_1^2}{2} \left[\left(\frac{A_1}{A_2} \right)^2 - 1 \right]$$

Substituting [24] in [25] and rearranging gives:

$$R_x = p_1 \left(A_1 - A_2 \right) + \frac{\rho v_1^2}{2} \left[2A_1 - A_2 \frac{A_1^2}{A_2^2} \right] \qquad [26]$$

If, as indicated previously, the dynamic head is relatively insignificant, this equation reduces to:

$$R_x = p_1 \left(A_1 - A_2 \right) \qquad [27]$$

A1.5 Forces on dead-ends

$$R_x = pA$$

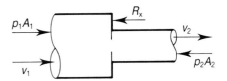

$$R_x = pA \qquad [28]$$

A1.6 Forces on valves and constrictions

From conservation of linear momentum:

$$\sum F_x = \rho Q \left(v_{2x} - v_{1x} \right) \quad +\rightarrow:$$

$$p_1 A_1 - p_2 A_2 - R_x = \rho Q \left(v_2 - v_1 \right)$$

or

$$R_x = p_1 A_1 - p_2 A_2 - \rho Q \left(v_2 - v_1 \right)$$

and since

$$A_1 v_1 = A_2 v_2$$

$$R_x = p_1 A_1 - p_2 A_2 - \rho A_1 v_1^2 \left[\frac{A_1}{A_2} - 1 \right] \qquad [29]$$

This case applies to restrictions to the flow, such as partially open valves in which there is a viscous head loss.

The pressure p_1, p_2 either side of the constriction can be related through the energy equation, however in this case friction losses cannot be ignored.

$$\text{i.e. } (p_2 - p_1) + \frac{\rho}{2}\left(v_2^2 - v_1^2\right) + \rho g h_L = 0 \qquad [30]$$

h_L represents the viscous losses and is normally related to the upstream flow velocity v_1 by a non-dimensional loss coefficient k, i.e.

$$h_L = k\frac{v_1^2}{2g} \qquad [31]$$

Combining [30] and [31] and rearranging yields:

$$p_2 = p_1 - \frac{\rho v_1^2}{2}\left[\left(\frac{A_1}{A_2}\right)^2 - 1\right] - k \cdot \frac{\rho v_1^2}{2} \qquad [32]$$

Substituting equations [32] into [29] and rearranging gives:

$$R_x = p_1\left(A_1 - A_2\right) + \frac{\rho v_1^2}{2}\left[2A_1 - \frac{A_1^2}{A_2} + k\,A_2\right] \qquad [33]$$

Practical interpretations of equations [29] and [33] are:

- equation [29], for a simple restriction with no change in cross-sectional area across the component, i.e. $A_1 = A_2 = A$, then, if p_1 and p_2 are both known,

$$R_x = (p_1 - p_2)\,A \qquad [34]$$

- if, on the other hand, the pressures are unknown but the flow rate, or velocity, is known and a value for the loss coefficient k can be estimated then from [33]

$$R_x = k\,\frac{\rho v_1^2}{2}\,A = K\,\frac{\rho Q^2}{2A} \qquad [35]$$

Appendix 2 Field identification of soils and rocks

Detailed identification and classification of soils and rocks for civil engineering purposes requires experience and detailed knowledge of soil mechanics, rocks mechanics and geology. For the purpose of routine design of thrust blocks, however, detailed classifications are not required and it is necessary only to identify the broad ranges of soils and rocks in Table 3.

These broad ranges depend on:

- grain size (fine or coarse)

- consistency

- presence and degree of cementing.

Correct identification of soils and rocks by visual inspection of hand samples is often quite difficult. There are, though, a number of simple procedures which may be followed to aid field identification of soils and rocks:

> NOTE: These procedures should be sufficient to identify the broad soil or rock type present for routine design of class 1 thrust blocks. If there is any doubt about the correct classification, expert advice should be sought.

1. **Take an intact lump of soil, about 25 mm to 50 mm cube, from the ground in the region of a thrust block.**

The sample may be taken from the side of the trench but loose and softened material should be removed to expose undisturbed ground. The sample may be taken with a spade or trowel or a thin wall tube (say 30 mm diameter) may be pushed or driven into the ground and a core recovered from inside the tube.

If it proves impossible to take an intact sample which holds together, the soil is probably an **uncemented sand or gravel**.

If it is impossible to take a sample because the ground is too hard it can probably be classified as **rock**.

2. **Place the intact sample in water (in a jam jar or a glass beaker). Try to avoid too much turbulence as the sample is lowered into the water.**

If the sample disintegrates and forms a rough cone, it is **uncemented silt, sand or gravel and $c' = 0$.**

It is then necessary to assess the consistency (see Table 3) to estimate the friction angle ϕ'.

If the sample stays intact for at least an hour then either it is a **fine-grained clayey soil or cemented sand or gravel or rock.**

3. **Try to remould the sample, still under water using a spoon or some similar implement or finger pressure.**

If the sample can be remoulded relatively easily, then it is **clay or weakly cemented sand or gravel.**

If the sample can only be remoulded with difficulty, then it is **moderately cemented sand or gravel or weak rock.**

If the sample still remains intact even after strong pressure, then it is **relatively strong rock.**

Ground classified as strongly cemented or as rock should be examined carefully to determine the closeness and directions of the joints and fissures (see Table 3).

4. **If the sample has been completely remoulded shake the soil and water vigorously so that the grains become a suspension. Then place the container on a level surface and allow the grains to sediment for about 1 hour.**

Examine the sediment in the glass container by eye and using a simple hand magnifying glass (say $\times$ 10).

Estimate the grading of the soil from the thicknesses of the sedimented layers: these will appear (from the top) as:

- organic material (peat) floating on the surface

- clay sized particles which will remain in suspension

- silt sized particles which are invisible to the naked eye (but can be seen through a hand lens)

- sand and gravel sized particles which can be seen with the naked eye.

If the soil is predominantly fine grained (over 35% clay) estimate the undrained shear strength s_u of the undisturbed soil in a freshly exposed trench side (see Table 3).

If the soil is predominately coarse grained estimate the consistency and the degree of cementing (see Table 3), using the guidelines below:

- if the submerged sample can just be handled without falling to pieces, it is lightly cemented and $c'<20$ kPa

- if the submerged sample withstands rough handling but can be broken down by strong finger pressure, it is moderately cemented.

- if the submerged sample cannot be broken by hand, it is strongly cemented and $c'>100$ kPa.

Appendix 3 Basic soil mechanics theory

A3.1 INTRODUCTION

Soils and rocks, like steel and concrete, have the basic characteristics of strength and stiffness. These determine the ultimate loads which can be resisted and the way in which ground movements develop as loads are increased. An understanding of these basic characteristics, particularly for soils, will be required for even simple thrust block designs.

Steel and concrete are essentially single-phase materials. Their stress-strain behaviour is assumed to be linear and elastic and, for a given grade of steel or concrete mix, the working strength has a single value. Soils are a little more complicated and have the following general properties:

1. They are generally two-phase materials consisting of mineral grains and water. (In unsaturated soils a gas — usually air or water vapour is present as well). The pressures in the fluid phase — the pore pressures — have a profound influence on soil behaviour.

2. Soils are highly compressible and volume changes occur largely due to rearrangement of the grains. As a result, the compressibility of soils is not linear or elastic and volume changes are related to the logarithm of stress.

3. In saturated soils, volume changes can only occur if water flows out of the soil, so ground movements will be delayed as it takes time for this to occur.

4. Soil strength is essentially frictional (i.e. the resistance to shear increases linearly with pressure). In addition, because water content or volume are related to pressure, soil strength is related to water content.

The following treatment of basic soil mechanics is highly simplified. The intention is to cover the fundamental theories of effective stress, compression and strength in a way which will explain the methods used in this report to calculate earth pressures on simple thrust blocks. The ideas presented are general and widely applicable to other ground engineering problems. They apply to many commonly found soils but it must be remembered that there are some unusual soils which act and react in ways outside of these theories. If it is suspected that the soils encountered are in this category then a geotechnical specialist should be consulted.

A3.2 CLASSIFICATION OF SOILS AND ROCKS

There are many schemes for classifying soils and rocks. These are usually based on simple descriptions of the appearance of the material and on the measurement of basic characteristics, like grain size and water content. Classifying a soil or rock will often give a good indication of its likely engineering behaviour. Classification essentially tries to describe the nature of a soil (i.e. what it consists of) and its state (i.e. how closely the grains are packed together) and whether there is any cementing between the grains.

A3.2.1 The nature of the material

The most important factor is the grain size, and the distribution of the different sizes. A scheme for classifying grain size is given in Table A3.1. Note that the grain sizes increase logarithmically from clay grains (<0.002 mm) which are very small to cobbles (>60 mm) which are relatively large. The most important consequence of grain size is the influence on permeability which, in turn, governs the rate at which water can flow through soil.

Table A3.1 *Particle size classification*

Description	Size (mm)
Cobbles	> 60
Gravel	60 to 2
Sand	2 to 0.06
Silt	0.06 to 0.002
Clay	< 0.002

The variation of grain size is called grading and is shown as a grading curve on a particle size distribution chart (see Figure A3.1). Soils which have a relatively small range of grain size (e.g. marine clay, estuarine silt) are poorly graded; mixed soils which contain a wide range of grain sizes (e.g. boulder clay) are well graded.

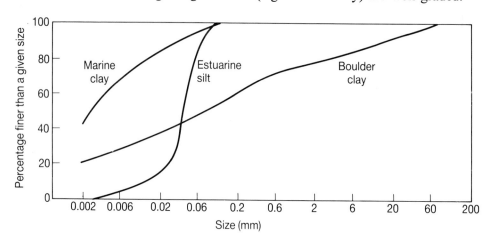

Figure A3.1 *Typical grading curves for soils*

As a very rough guide to particle sizes, sand and coarser grains can be seen with the naked eye, silt grains can be seen with a hand lens or optical microscope but clay grains can only be seen at very high magnifications in electron microscopes. There are various other simple indicators to grain size; for example, clay will stick to hands or boots but silt can be easily washed off or, if dry, dusted off.

For coarse-grained soils the engineering behaviour is governed largely by the grain size and grading. In fine grained clay soils, or in mixed soils with more than about 30 to 40% clay, the mineralogy of the clay has an important influence on behaviour. The mineralogy of a clay soil is described by its Atterberg limits — these are the liquid limit (LL), the plastic limit (PL) and the plasticity index (PI = LL − PL). The Atterberg limits of some common soils are given in Table A3.2.

Table A3.2 *Atterberg limits of some common soils*

Soil	LL	PL	PI
Boulder clay	30	15	15
Kaolin clay	65	35	30
London Clay	75	25	50

The liquid limit corresponds to the water content (in percentage) of a very soft recently deposited soil, while the plastic limit corresponds to the water content of heavily loaded, or heavily overconsolidated soil, and taken together they represent roughly the maximum and minimum water contents that a particular soil could have in the ground. Thus the plasticity index (PI) corresponds to the compressibility of the soil, i.e. the difference between the maximum and minimum water contents.

A3.2.2 The state of the material

It is well established that both the strength and the stiffness of a given soil vary with depth (i.e. with stress) and with water content. These two parameters — stress and water content — describe the state of the soil. For thrust blocks and relatively near surface structures, water content is the most important.

A very useful parameter for characterising the state of clay soils is the liquidity index $LI = (w - PL)/PI$, so that $LI = 1$ when the water content (w) equals the liquid limit and $LI = 0$ when the water content equals the plastic limit.

In coarse-grained soils, for which the Atterberg limits do not apply, the relevant parameters are the minimum density D_{min} (corresponding to the loosest state), the maximum density D_{max} (corresponding to the densest state) and the relative density D_r which ranges from zero at the minimum to 1.0 at the maximum density. The relative density of a granular soil is equivalent to the liquidity index for clay soils (except the numbers are reversed; i.e. $LI = 1 - D_r$ so that $D_r = 0$ is equivalent to $LI = 1.0$).

Figure A3.2(a) illustrates the variation of liquidity index with stress for a typical clay soil. During deposition the liquidity index decreases with increasing stress (i.e. with increasing depth of burial) but if soil is eroded so that the stress decreases the liquidity index remains approximately constant. Figure A3.2(b) illustrates the corresponding variation of liquidity index with depth for freshly deposited soils and for soils where material has been eroded from above. Freshly deposited soils will be relatively soft and weak while pre-loaded soils will be relatively stiff and strong because strength and stiffness change with water content.

Coarse-grained soils with low relative density (i.e. relatively loose) will be soft and weak, while those with high relative density (i.e. relatively dense) will be stiff and strong.

The nature of a soil, given by grading and mineralogy, and its state, given by water content, liquidity index or relative density, are the most important factors in classifying soils as, taken together, they indicate the relative magnitudes of the expected strength and stiffness. A number of other factors, such as the presence of peat or cementing, will influence the behaviour of soils. Rocks are generally relatively stiff and strong, compared to soils, although the presence of joints and fissures can substantially influence their engineering properties.

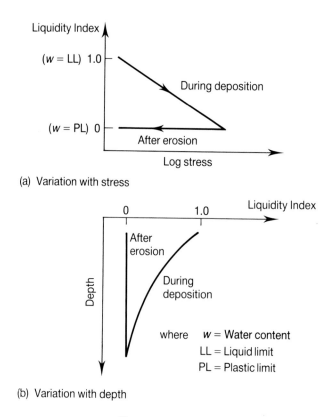

(a) Variation with stress

(b) Variation with depth

Figure A3.2 *Variation of liquidity index with stress and depth for normally consolidated and overconsolidated soils*

Often it is helpful to give the geological name of a soil or rock (i.e. London Clay, Greensand) as geotechnical engineers will know from their experience the likely engineering properties of local geological strata. These geological descriptions should, however, be used with caution as, even within a particular stratum, the properties of the materials will vary and they will also be influenced by things like loading and weathering.

Full schemes for describing soils and rocks are given in BS 5930 1981[5] which is the British Standard for site investigations. The present, very brief, treatment should be related to the information given in Table 3 in this Report and also to Appendix 2 which describes simple procedures for field identification of soils and rocks.

A3.3 STRESS AND EFFECTIVE STRESS

Stresses in soils and rocks must take account of the pore pressures in the fluid phase. Total vertical stresses σ_v, on elements at the same depth, Z, are illustrated in Figure A3.3. Here, γ is the unit weight of the soil (usually $\gamma = 17$ to 21 kN/m³. for saturated soil) and $\gamma_w = 9.81$ kN/m³ is the unit weight of water. The total stress is the stress due to everything — soil, water and surface loads — above the element. Pore pressures are given by the height of water, h_w, in a standpipe (when there is no flow into or out of the standpipe) as illustrated in Figure A3.4.

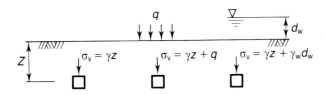

Figure A3.3 *Calculation of total vertical stress*

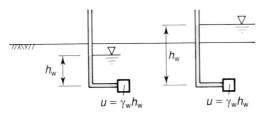

where u = Pore pressure

Figure A3.4 *Calculation of pore pressure*

The stress which governs soil behaviour is known as the effective stress, denoted σ', and given simply by the difference between the total stress and the pore pressure, so that

$$\sigma' = \sigma - u \qquad \text{[A3.1]}$$

The principle of effective stress — that all soil behaviour such as strength and compression are governed by the effective stresses given by equation A3.1 — is absolutely fundamental to soil mechanics. So far as we know, all soils obey the principle of effective stress, at least within the common range of stresses. From Figures A3.3 and A3.4 it follows that, if the water level is above ground level, a change of water level does not change the effective stress so the effective stresses 1 m below the bed of the deep ocean will be the same as the effective stresses 1 m below the bottom of a duck pond. On the other hand, if the water level is below ground level a change of water level will change the effective stresses.

The influence of effective stress is demonstrated in Figure A3.5 which shows settlement $\Delta\rho$ due to construction of a foundation and due to groundwater lowering. In the first case, in Figure A3.5(a), the pore pressures do not change so the change of effective stress is equal to Δq and this causes the settlements. In the second case in Figure A3.5(b) the total stresses do not change so the change in effective stress is equal to Δu (where Δu is negative because the pore pressures are reduced) and this causes the settlements. For an equal change in effective stress the settlements will be the same and it does not matter whether they were caused by changes of total stress, or of pore pressure, or both.

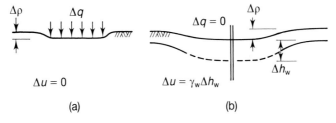

Figure A3.5 *Settlements caused by foundation loading and groundwater lowering*

While Figure A3.3 shows only the vertical stresses which act on horizontal planes there will, of course, be other stresses acting on other planes. Figure A3.6(a) shows a region of soil in front of a loaded thrust block and a triangular element of soil. The vertical and horizontal stresses are principal stresses so the shear stresses on horizontal and vertical planes are zero. There are shear and normal stresses τ_n and σ_n, on a plane inclined at an angle α. The corresponding Mohr's circle of total stress is shown in Figure A3.6(b). This shows the relationships between the principal stresses and the shear and normal stresses on the inclined plane.

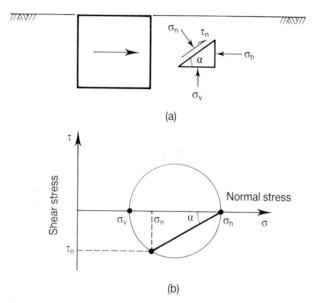

(a)

(b)

Figure A3.6 *Analysis of stress using the Mohr's Circle construction*

In Figure A3.6 the shear stress, τ, developed in the soil contributes resistance to the thrust block, which will remain stable provided that the mobilised shear stresses do not exceed the limiting shear stress which the soil can sustain. Thus, soil strength is simply the maximum shear stress that can be applied but, as will be seen later, this depends on the normal stress and on the water content.

Figure A3.7 shows Mohr's circles of total stress and of effective stress plotted on the same diagram. Since equation A3.1 applies for both the horizontal and vertical stresses, the Mohr's circles have the same diameter and the effective stress circle is simply moved to the left of the total stress circle by a distance equal to the pore pressure u. The points A and A' represent total and effective stresses on the same plane and from the diagram, the total and effective shear stresses are equal.

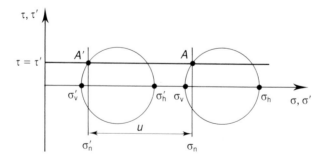

Figure A3.7 *Mohr's circles of total and effective stress*

A3.4 RATES OF LOADING AND DRAINAGE

As soil is loaded or unloaded, in compression or in shear, volume changes will occur as soil particles move relative to each other. Volume changes can only occur if water flows from, or into, the soil because the soil grains and (in saturated soils) the pore fluid are incompressible. If the loading is relatively slow there will be plenty of time for this flow to take place and the pore pressure will remain constant; conversely, if the loading is relatively fast there will not be enough time for flow to occur and the pore pressures will change.

Figure A3.8 shows the changes of stress, pore pressure and volume with time when the total stress is increased from, σ_o, by an increment, $\Delta\sigma$. In Figure A3.8(a) the load is applied slowly, over a long period of time, and the volume reduces from V_o by ΔV while the pore pressure remains constant at u_o. This is known as drained loading as all the drainage and volume changes take place during loading so that when the load is held constant the volume also remains constant. In this instance, $\Delta u = 0$ and $\Delta\sigma = \Delta\sigma'$ and the volume change, ΔV is governed by the change of effective stress $\Delta\sigma'$.

Figure A3.8(b) illustrates what happens when the load is applied quickly, over a short period of time. There is no time for any drainage and the volume remains unchanged at V_o but the pore pressure rises. The pore pressure near the load is now increased above the original pore pressure u_o by an excess pore pressure $\bar{u}$. This loading is known as undrained as there is no drainage during the loading. In time, water flows from the loaded soil driven by the excess pore pressures, so volume changes occur as the excess pore pressures drop and effective stresses rise. At some time t the excess pore pressure is $\bar{u}_t$, the change of effective stress is $\Delta\sigma'$ and the change of volume is ΔV.

The magnitudes of the excess pore pressures determine the rate of change of volume so that the rates of change of pore pressure and volume diminish with time. This process is known as consolidation and it accounts for most of the time dependent behaviour in soils. After a long time, the excess pore pressures have dropped to zero and the pore pressures, effective stresses and volume are the same as those at the end of the drained loading in Figure A3.8(a).

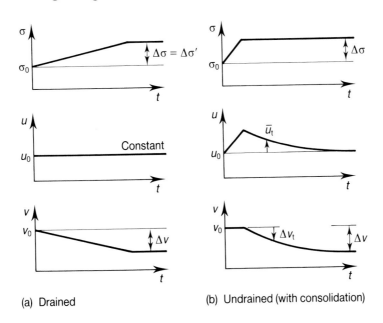

(a) Drained   (b) Undrained (with consolidation)

Figure A3.8 *Behaviour of soils during drained and undrained loading*

There is a mathematical theory for consolidation for a number of relatively simple cases but it is not easy to calculate pore pressures and effective stresses after undrained loading and during consolidation. For cases of drained loading, pore pressures do not change and so calculations can be carried out in terms of effective stresses. For cases of undrained loading, in general, it is not possible to determine effective stresses and so alternative calculations will be necessary. Consequently, it is important to determine whether a particular loading is essentially drained or undrained.

The important factors are the relative rates of drainage and loading. If the rate of loading is relatively slow compared with the rate of drainage the loading will be essentially drained, but if the rate of loading is relatively quick compared with the rate of drainage the loading will be essentially undrained and constant volume. Any undrained loading will be followed by consolidation as the excess pore pressures dissipate.

Flow of water in soils is governed by Darcy's law:

$$V = ki \qquad\qquad\qquad [A3.2]$$

in which V is the water flow velocity, i is the hydraulic gradient and k is the coefficient of permeability. Approximate values for k for typical soils are given in Table A3.3. The important points are that there is a very large range − more than six orders of magnitude − of k in natural soils, and grain size has a major influence on rate of drainage. Table A3.4 lists the approximate durations of typical engineering loadings. Again, there is a very wide range − more than nine orders of magnitude − between quick shock loads and slow loading due to natural erosion.

In determining whether a particular loading should be considered to be drained or undrained it is the relative rates of loading and drainage that matter. Therefore, even in a sand or gravel, shock loadings due to pile driving or earthquakes will be essentially undrained while, in a clay soil natural slopes will be essentially drained. It is not sufficient to classify all fine grained soils as undrained and all coarse grained soils as drained.

The design condition for thrust blocks is complicated. Thrust loads due to steady pipeline flow act for long periods of time and the soil will be drained in the life of the pipeline, while transient loads due to valve closure etc. are relatively rapid and most soils would be undrained during these transient loadings. For the purposes of this report, and to provide a relatively simple but economic design procedure, thrust blocks will be considered to apply drained loading in sands, gravels and silts and to apply undrained loading in fine-grained clay soils. In the latter case some modifications will be made to take account of softening with time.

Table A3.3 *Rates of drainage for typical soils*

Soil	Coefficient of permeability k m/s
Gravel	$> 10^{-2}$
Sand	$10^{-2} - 10^{-5}$
Silt	$10^{-5} - 10^{-8}$
Clay	$< 10^{-8}$

Table A3.4 *Duration of typical loadings*

Event	Duration(s)
Shock	1
Wave	10
Trench excavation	10^4 (3 hours)
Small foundation	10^7 (3 months)
Embankment	10^8 (3 years)
Erosion	10^9 (10 years)

A3.5 COMPRESSION AND SWELLING OF SOILS

Soil will compress or swell, reacting to any increase or decrease in effective normal stresses σ'_n, as the grains rearrange themselves and the water content changes. A reduction in water content provides less opportunity for volume changes due to further particle rearrangement, resulting in a non-linear stress-strain curve and an increase in bulk modulus with loading. Conversely, unloading provides less opportunity for particle rearrangement, so that the stiffness in unloading will be larger than in loading.

Figure A3.9(a) illustrates a typical volumetric stress-strain curve for soil with an unloading and reloading loop. Under isotropic loading conditions the gradient of the curve gives the bulk modulus:

$$K' = \frac{d\sigma'}{d\varepsilon_v} \qquad \qquad [A3.3]$$

where K' depends on the current stress and on whether the soil is being loaded or unloaded.

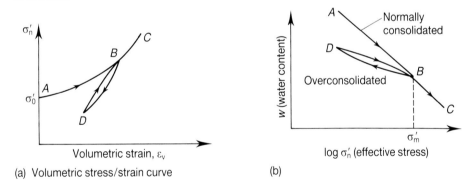

(a) Volumetric stress/strain curve   (b)

Figure A3.9 *Compression and swelling of soil*

Compression and swelling of soil is usually shown as water content plotted against the logarithm of effective stress as illustrated in Figure A3.9(b). Now the non-linear stress-strain curves in Figure A3.9(a) become approximately linear and may be written as:

$$w = w_0 - C\log\sigma'_n \qquad \qquad [A3.4]$$

where w_0 is the water content at $\sigma' = 1$kPa and C is either a coefficient of compression, C_c, or a coefficient of swelling, C_s. Values for C_c and C_s for some typical soils are given in Table A3.5.

Table A3.5 *Coefficients of compression and swelling for some common soils*

Soil	C_c	C_s
Boulder clay	0.1	0.02
Kaolin clay	0.5	0.1
London clay	0.4	0.1

During first loading, along AB and along BC in Figure A3.9(a) the soil is known as normally consolidated and ABC is the normal consolidation line. During unloading and reloading on the loop BD the soil is known as overconsolidated. If the current state of stress lies along BD is σ' then the overconsolidation ratio R_o is given by:

$$R_o = \frac{\sigma'}{\sigma'_m}$$

[A3.5]

where σ'_m is the stress at B at the intersection of the unloading-reloading loop and the normal consolidation line.

For normally consolidated soils $R_o = 1$ and the overconsolidation ratio increases with erosion or with unloading.

The compression and swelling behaviour illustrated in Figure A3.9 is typical for fine-grained and mixed soils. The behaviour of granular soils, illustrated in Figure A3.10 is similar except that the normal consolidation line, if one exists, is far to the right of the diagram at very high stresses that are not usually reached in engineering structures. The behaviour of loose and dense samples follow loading and unloading loops similar to those in Figure A3.9. Dense soils are stiffer than the loose soils because they are further from the possible normal consolidation line i.e. they are apparently more heavily overconsolidated.

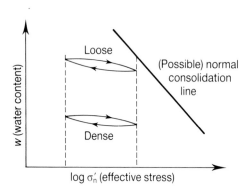

Figure A3.10 *Compression and swelling of granular soils*

Compression and swelling of soils is clearly important in determining the settlement of foundations and the heave of excavations but it is less important for the design of thrust blocks subjected to horizontal loads. The important features illustrated in Figure A3.9 are the non-linear stress-strain behaviour and the distinction between normally consolidated and overconsolidated soils: the former will be relatively soft (i.e. they will have relatively low bulk modulus) while the latter will be considerably stiffer.

A3.6 STRENGTH OF SOILS

The strength of a soil (or of any material) is described by the maximum shear stress it can sustain and, for soil, this depends on the normal effective stress and the water content and on the strain.

Figure A3.11(a) shows a block of drained soil subjected to a normal effective stress σ'_n and an increasing shear stress. This might be in a laboratory test or in the ground near a thrust block (see Figure A3.6). Figure A3.11(b) shows typical shear stress-strain curves for overconsolidated or dense soils and for normally consolidated or loose soils, both with the same effective normal stress. Figure A3.11(c) shows the corresponding volumetric strains.

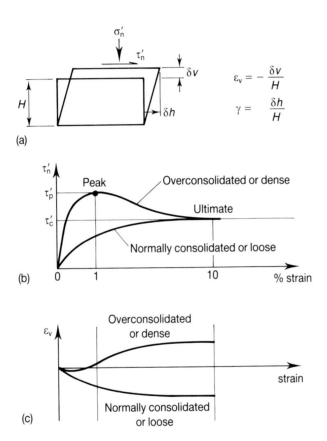

Figure A3.11 *Behaviour of soils during shear*

Normally consolidated or loose soil compresses as the water content reduces and the shear stress gradually increases until it reaches an ultimate state where the shear stress reaches a critical value τ'_c. Overconsolidated or dense soil dilates (after a small initial compression) and there is a peak shear stress τ'_p. Beyond the peak the increasing water content due to the dilation causes the shear stress to drop until the soil reaches the same critical value τ'_c. This characteristic behaviour, illustrated in Figure A3.11, is typical of almost all uncemented soils in any kind of loading, in the laboratory or in situ, in which shear stresses increase.

During undrained loading the volumetric strains are zero (by definition) but the pore pressures will change. These changes reflect the volumetric strains in a corresponding drained test: in normally consolidated or loose soils the pore pressures will rise and there will be positive excess pore pressure, whereas in overconsolidated or dense soils the pore pressures will fall and there will be negative excess pore pressures. During undrained loading the water content remains constant so the soil will not dilate and soften after the peak and the shear stresses at the peak and ultimate states will be approximately the same.

If soil is loaded undrained to states below failure and the load is maintained, as is normal in engineering works, the excess pore pressures dissipate with time. In normally consolidated or loose soils the positive excess pore pressures drop so effective stresses increase and the soil strengthens. In overconsolidated or dense soils the negative excess pore pressures rise, so effective stresses fall and the soil softens and weakens.

If a number of tests are carried out on samples of the same soil with different effective normal stresses and different overconsolidation ratios the peak and ultimate states will appear as illustrated in Figure A3.12. The ultimate states will fall on unique lines given by:

$$\tau'_c = \sigma'_n \tan\phi'_c \qquad\qquad\qquad [A3.6]$$

and $\quad w_c = w_{co} - C_c \log\sigma'_n \qquad\qquad\qquad [A3.7]$

where ϕ'_c is the critical friction angle, C_c is the coefficient of compression (equation A3.4) and w_{co} is the water content of a sample at its ultimate state with $\sigma'_n = 1.0$ kPa.5

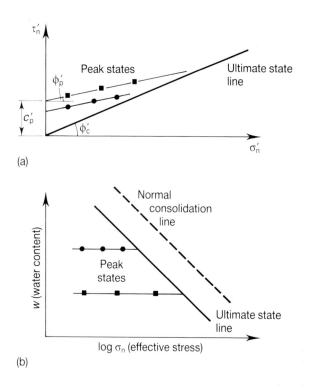

(a)

(b)

Figure A3.12 *Peak and ultimate failure states*

Plotting w against $\log\sigma'_n$ the ultimate state and normal consolidation lines are straight and parallel and have the same gradient, equal to the consolidation coefficient, C_c.

The peak states fall on a set of lines given by:

$$\tau'_p = c'_p + \sigma'_n \tan\phi'_p \qquad [A3.8]$$

where ϕ'_p is the peak friction angle and c'_p is the peak cohesion intercept.

Each peak state line corresponds to a particular water content as shown in Figure A3.12(b), and so c'_p depends on the water content. It should be noted that the peak cohesion intercept, c'_p, occurs even in reconsolidated clays and clean granular soils. The peak cohesion intercept is, therefore, primarily the result of dilation (see Figure A3.11) and any cementing or inter-particle bonding may lead to an additional contribution to strength.

There is an alternative formulation for peak state which is often used, especially for sands and gravels. Figure A3.13(a) illustrates shear stress-strain curves for four samples of the same soil tested at different relative densities but with the same normal effective stress – sample A is loose and sample D is dense, and both peak states can be represented by:

$$\tau'_p = \sigma'_n \tan\phi'_m \qquad [A3.9]$$

where the angle ϕ'_m is shown in Figure A3.13(b). For very loose samples (relative density approximately zero) the value of ϕ'_m will be close to the ultimate friction angle ϕ'_c and ϕ'_m will increase with increasing relative density as illustrated in Figure A3.13(c). Table A3.6 gives characteristic valves for ϕ'_c and ϕ'_m.

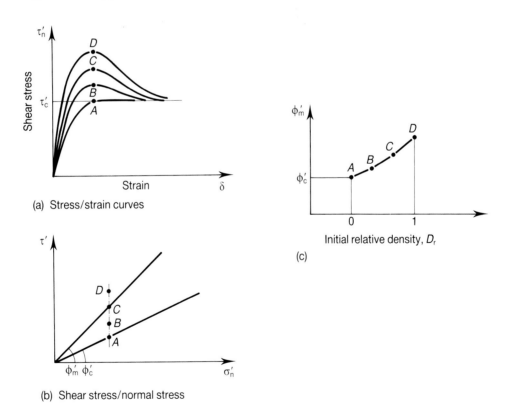

Figure A3.13 *Peak states for loose and dense soils*

Table A3.6 *Friction angles for some common soils*

Soil	ϕ'_c	ϕ'_m	(at max relative density)
Carbonate sand	38°	44°	
Beach sand	30°	37°	
Boulder clay	28°	—	
London Clay	22°	—	

Equations A3.6 and A3.9 give the ultimate and peak state strengths of soils in terms of the normal effective stress. This can be found only if the pore pressures are known, which will generally be the case only for drained loading. Under undrained loading, the pore pressures will usually be unknown and alternative methods are required to determine the undrained strength of soils.

Figures A3.14(a) and A3.14(b) show ultimate state lines and these are the same as those shown in Figure A3.12. Figure A3.14(c) shows the same ultimate state line plotted as shear stress against water content and this demonstrates that the ultimate strength of soil is uniquely related to its water content. Noting that $\tau_n = \tau'_n$, the undrained strength

$$\tau_c = s_u \qquad\qquad [A3.10]$$

is determined by the corresponding water content w_c. Since water content remains constant during undrained loading the undrained strength used in a design is the strength measured in any field or laboratory test on samples at the in-situ water content. Of course, if the water content changes the strength will change as shown by Figure A3.14(c) − a rise in water content will lead to a reduction in strength.

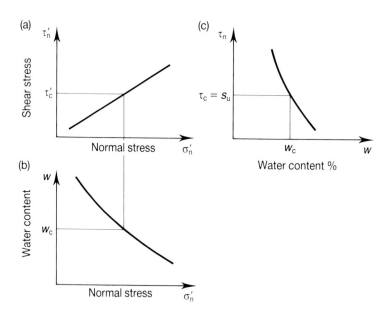

Figure A3.14 *Ultimate failure states for undrained loading*

To assess the stability of thrust blocks, the strength of uncemented soils can be represented by equation A3.9 (for drained loading or coarse grained soils) and by equation A3.10 for fine grained clay soils. Cemented granular soils require the addition of cohesion in equation A3.10 to account for the additional strength.

The assumption of undrained loading requires that the water content of the soil on the passive side of the thrust block does not increase — this is reasonable as horizontal loads transferred through the thrust block into the ground will tend to reduce the water content. On the active side of the thrust block the stress will be reduced and the soil will soften as water content increases. Safe design of thrust blocks in clay soils assumes that soil on the active side may soften fully and have zero strength.

A3.7 STIFFNESS OF SOILS

Soils, like other materials, have the properties of strength and stiffness. Strength is the maximum shear stress that can be sustained by the soil, and depends on the normal effective stress σ'_n. Stiffness is the relationship between changes of stress and strain. These are not uniquely related and, as illustrated in Figure A3.15, materials may have any combination of strength (strong or weak) and stiffness (stiff or soft).

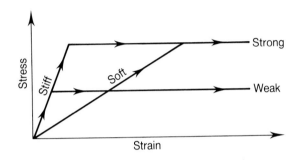

Figure A3.15 *Strength and stiffness*

Figure A3.16(a) shows idealised stress-strain curves for dense or overconsolidated samples and loose or normally consolidated samples of the same soil. The shear stresses have been normalised by dividing by the normal effective stress σ'_n so both curves reach the same ultimate state given by equation A3.6. The shear modulus G' is the gradient of the stress-strain curve, such that:

$$G' = \frac{d\tau}{d\gamma} \qquad \text{[A3.11]}$$

and G'_o is the initial stiffness. The initial stiffness, normalised by dividing by the normal effective stress, increases with increasing relative density, or with increasing overconsolidation, as illustrated in Figure A3.16(b).

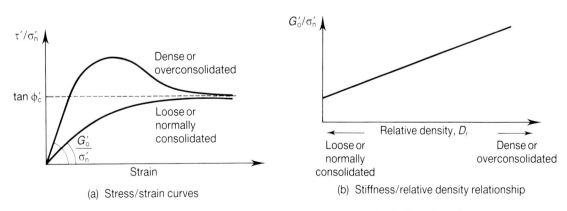

(a) Stress/strain curves

(b) Stiffness/relative density relationship

Figure A3.16 *Relationships between initial stiffness and relative density*

Thrust blocks must be designed primarily to limit ground movements and the applied loads must be considerably less than those to cause failure. Soil stiffness determines ground movements and depends on relative density (see Figure A3.16(b)). Therefore thrust blocks in dense or overconsolidated soils may have proportionally greater loads (i.e. lower load factors) than thrust blocks in loose or normally consolidated soils.

Appendix 4 Derivation of formula for ground resistance

A4.1 INTRODUCTION

The calculations used to determine the ultimate ground resistance of a thrust block are those from basic soil mechanics theory for active and passive earth pressures and for bearing capacity. (These are covered in detail in many texts on soil mechanics and foundation engineering).

The ground resistance will come from:

1. Active and passive earth pressures acting on the front and back faces

2. Shear stresses acting on the base

3. Bearing pressures acting on the base

4. The weight of the block and the soil above it.

As with all calculations in soil mechanics it is necessary to distinguish the two cases:

1. For undrained conditions where the analyses are carried out in terms of total stresses using the undrained strength s_u.

2. For drained conditions where the analyses are carried out in terms of effective stresses using the effective stress strength parameters c' and ϕ'.

Undrained analyses will normally be appropriate for fine-grained soils and intact rocks and drained analyses will normally be appropriate for coarse-grained soils and jointed rocks.

A4.2 EARTH PRESSURES FOR UNDRAINED CONDITIONS

Figure A4.1(a) shows elements of soil P and A on the passive and active sides of a thrust block. Figure A4.1(b) shows the Mohr's circles of total stress for these elements for the case where the undrained strength on both sides is the same. Due to the loads on the block, the soil on the passive side may consolidate a little and the strength will increase as the water content decreases. However, the soil on the active side will be unloaded, and will soften and swell, and the strength will reduce as the water content increases. For simplicity and safety it should be assumed that soil on the active side softens so that the strength is almost zero, while the soil on the passive side has a negligible gain in strength.

The corresponding Mohr's circles are shown in Figure A4.1(c), the fully softened strength should be take as zero so that $\sigma_a = \sigma_v$. From the geometry of Figure A4.1(c) we have:

$$\sigma_p - \sigma_v = 2s_u \qquad \text{[A4.1]}$$

$$\sigma_a = \sigma_v \qquad \text{[A4.2]}$$

and hence

$$(\sigma_p - \sigma_a) = 2s_u \qquad \text{[A4.3]}$$

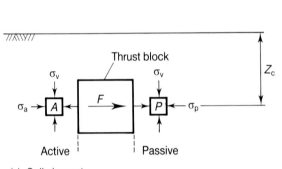

(a) Soil elements

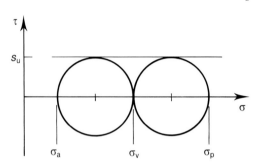

(b) Mohrs circles for undrained strength same for both active and passive sides

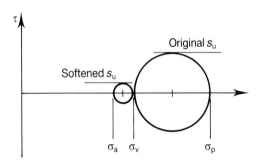

(c) Mohrs circles for active side softened (s_u tends to zero) and passive side undrained

Figure A4.1 *Earth pressures for undrained loading*

A4.3 EARTH PRESSURES FOR DRAINED CONDITIONS

Figure A4.2(a) shows elements of soil P and A on the passive and active sides of a thrust block. The stresses on the elements are total stresses. These are the sum of the effective stresses and the pore pressures and so:

$$\sigma = \sigma' + u \qquad \text{[A4.4]}$$

Figure A4.2(b) shows the Mohr's circles of *effective* stress for these elements for a cohesionless soil with $c' = 0$. From the geometry of Figure A4.2(b) we have

$$\sigma'_p = \sigma'_v K_p \qquad \text{[A4.5]}$$

$$\sigma'_a = \sigma'_v K_a \qquad \text{[A4.6]}$$

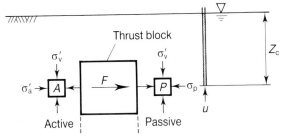

(a) Soil elements

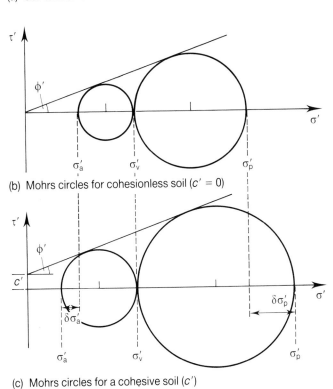

(b) Mohrs circles for cohesionless soil ($c' = 0$)

(c) Mohrs circles for a cohesive soil (c')

Figure A4.2 *Earth pressures for drained loading*

where K_p and K_a are the conventional Rankine earth pressure coefficients given by:

$$K_p = \tan^2 \left(45° + \frac{\phi'}{2} \right) = \frac{1 + \sin\phi'}{1 - \sin\phi'} \qquad \text{[A4.7]}$$

$$K_a = \tan^2 \left(45° - \frac{\phi'}{2} \right) = \frac{1 - \sin\phi'}{1 + \sin\phi'} \qquad \text{[A4.8]}$$

For drained conditions the pore pressures on the passive and active sides of the thrust block are the same and hence, we have:

$$(\sigma_p - \sigma_a) = (\sigma_p' - \sigma_a') \qquad \text{[A4.9]}$$

$$= \sigma'_v (K_p - K_a) \qquad \text{[A4.10]}$$

where σ'_v is the *effective* vertical stress at the depth Z_c and is given by:

$$\sigma'_v = \gamma\, Z_c - u \qquad\qquad\qquad\qquad [A4.11]$$

where γ is the total unit weight of the soil. When the groundwater table is at the ground surface, as in Figure A4.2(a) we have:

$$u = \gamma_w\, Z_c \qquad\qquad\qquad\qquad [A4.12]$$

and, from the above,

$$(\sigma_p - \sigma_a) = (\gamma - \gamma_w)\, Z_c\, (K_p - K_a) \qquad\qquad\qquad\qquad [A4.13]$$

If the water table is below the level of the base of the thrust block $u = 0$ or $\gamma_w = 0$.

Figure A4.2(c) shows Mohr's circles of *effective* stress for elements P and A for the case where the soil is cemented with a cohesion intercept c'.

From the geometry of the figure we have:

$$\delta\sigma'_p = 2c'\,\tan\left(45° + \frac{\phi'}{2}\right) = c'\,K_{pc} \qquad\qquad\qquad\qquad [A4.14]$$

$$\delta\sigma'_a = 2c'\,\tan\left(45° + \frac{\phi'}{2}\right) = c'\,K_{ac} \qquad\qquad\qquad\qquad [A4.15]$$

where $K_{pc} = 2\sqrt{K_p}$ and $K_{ac} = 2\sqrt{K_a}$

Thus, we have

$$\sigma'_p = \sigma'_v K_p + c'\,K'_{pc} \qquad\qquad\qquad\qquad [A4.16]$$

$$\sigma'_a = \sigma'_v\, K_o - c'\,K_{ac} \qquad\qquad\qquad\qquad [A4.17]$$

and hence

$$(\sigma_p - \sigma_a) = (\gamma - \gamma_w)\, Z_c\, (K_p - K_a) + c'(K_{pa} + K_{ac}) \qquad\qquad\qquad\qquad [A4.18]$$

A4.4 BASE SHEARING RESISTANCE

Some resistance to sliding will be developed as shear stresses between the base of the thrust block and the soil (see Figure A4.3(a)). In order to mobilise this shearing resistance it is necessary to ensure that there is a rough contact between the concrete of the thrust block and the base of the excavation which should be artificially roughened. Alternatively, full base shear resistance can be mobilised by constructing a step or key (see Figure A4.3(b)).

NOTE: For downturn bends there are additional vertical (upward) forces which will reduce the normal stresses on the base. In this case, base shear stresses should be reduced or neglected.

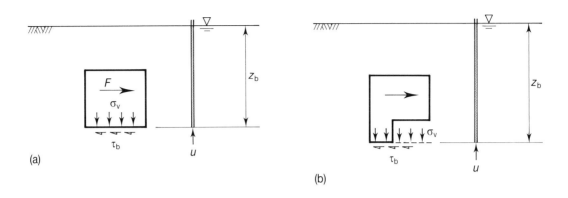

Figure A4.3 *Base shearing resistance (a) Development of base shear stress (b) Use of key to mobilise base shear resistance*

A4.4.1 Base resistance for undrained conditions

For undrained conditions the shear stress on a roughened base is equal to the undrained strength of the soil and hence:

$$\tau_b = s_u \qquad [A4.19]$$

A4.4.2 Base resistance for drained conditions

The base shear stresses for drained conditions are given by the basic soil strength equation. Hence, noting that total and effective shear stresses are always equal for any form of loading we have, for cohesionless soil:

$$\tau_b = \sigma_v' \tan\phi' \qquad [A4.20]$$

where σ_v' is the vertical *effective* stress at the depth Z_b and is given by

$$\sigma_v' = \gamma Z_b - u \qquad [A4.21]$$

where γ is the unit weight of the soil.

Where the groundwater table is at the ground surface, as in Figure A4.3 we have

$$u = \gamma_w Z_b \qquad [A4.22]$$

and, from the above,

$$\tau_b = (\gamma - \gamma_w)Z_b \tan\phi' \qquad [A4.23]$$

If the water table is below the level of the base of the block $u = 0$ or $\gamma_w = 0$.

For cemented soils which have a cohesion intercept, c', the basic strength equation becomes

$$\tau_b = c' + \sigma_v' \tan\phi' \qquad\qquad [A4.24]$$

and hence

$$\tau_b = c' + (\gamma - \gamma_w)Z_b \tan\phi' \qquad\qquad [A4.25]$$

A4.5 BEARING CAPACITY FOR UPTURN BENDS

For upturn bends the total bearing pressures on the base arise from the *sum* of the vertical component of the fluid force, F_v, plus the weight of the block and soil above it, Figure A4.4. The bearing pressure of the soil, q_b, is required to resist the additional forces due to fluid pressures with small settlements.

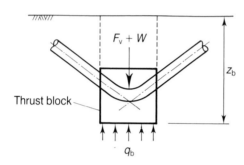

Figure A4.4 Bearing capacity for upturn bends

A4.5.1 Bearing capacity for undrained conditions

The total ultimate bearing capacity of a foundation for undrained conditions is given by

$$q_b = s_u N_c + \gamma Z_b \qquad\qquad [A4.26]$$

where N_c is a bearing capacity factor and γZ_b is the total vertical stress at the base level before construction. Since the unit weights of soil and concrete are approximately equal the weight of the thrust block and the soil above it apply a total vertical stress approximately equal to γZ_b. Hence, the bearing capacity available to resist the fluid forces is:

$$q_b = s_u N_c \qquad\qquad [A4.27]$$

For thrust blocks which are approximately square in plan the appropriate value of N_c is about 6 and hence:

$$q_b = 6s_u \qquad\qquad [A4.28]$$

A4.5.2 Bearing capacity for drained conditions

The total ultimate bearing capacity of a thrust block for drained conditions in soil or rock with a friction angle ϕ' and a cohesion intercept c' is given by:

$$q_b = c' \, N_c + \tfrac{1}{2} \, (\gamma - \gamma_w) B \, N_\gamma + (\gamma - \gamma_w)(N_q - 1)Z_b + \gamma \, Z_b \qquad \text{[A4.29]}$$

where N_c, N_γ and N_q are bearing capacity factors, B is the minimum width of the thrust block, and $\gamma \, Z_b$ is the total vertical stress at the base level before construction. As before, since the unit weights of soil and concrete are approximately equal, the weight of the thrust block and soil above it apply a total vertical stress approximately equal to $\gamma \, Z_b$. Hence, the bearing capacity available to resist the fluid force is:

$$q_b = c' \, N_c + \tfrac{1}{2} \, (\gamma - \gamma_w) B \, N_\gamma + (\gamma - \gamma_w)(N_q - 1)Z_b \qquad \text{[A4.30]}$$

Values for the bearing capacity factors for values of ϕ' in the range 20 to 35 are given in Table 6: intermediate values may be obtained by interpolation from the table.

For soil with no strength ($c' = 0$ and $\phi' = 0$) and with unit weight $\gamma = \gamma_w$ the bearing capacity equation should reduce to the expression for a body floating in water. Multiplying the bearing capacity equations by the base area A_b and with $N_q = 1, N_c = 0$ and $N_\gamma = 0$ we have:

$$Q = q_b \, A_b = \gamma_w Z_b \, A_b \qquad \text{[A4.31]}$$

This says that the weight Q of the floating body is equal to the weight $\gamma_w \, Z_b \, A_b$ of fluid displaced which is, of course, Archimedes Principle.

NOTES:

1. The active and passive pressure formulae are for thrust blocks with smooth faces where there is no shear stress between the faces of the blocks and the soil. There are a number of alternative formulae for rough faced blocks where shear stresses can be mobilised between the soil and the faces of the block. Use of these alternative formulae result in higher resistances and they should be used with caution.

2. The methods and formulae for calculating soil resistances apply for cases where the lines of action of the applied fluid forces and the ground resistance are approximately coincident so that the stresses are approximately uniform and there are no major moments or rotations.

Core Programme Members (November 1994)

Consultants

ACER Consultants Ltd
Ove Arup Partnership
Aspinwall & Co
W S Atkins Consultants Limited
Babtie Group Ltd
Binnie & Partners
Building Design Partnership
Curtins Engineering Consultants plc
Davis Langdon & Everest
Mark Dyer Associates
Sir Alexander Gibb & Partners Limited
Graham Consulting Group
Sir William Halcrow & Partners Ltd
G Maunsell & Partners
Montgomery Watson Ltd
Mott MacDonald Group Ltd
L G Mouchel & Partners Ltd
Posford Duvivier
Rendel Palmer & Tritton
Rofe Kennard & Lapworth
Scott Wilson Kirkpatrick & Partners
Thorburn Colquhoun
Wardell Armstrong
Sir Owen Williams and Partners
Geotechnical Ltd

Water Utilities

Northumbrian Water Limited
North West Water Limited
South West Water Services Ltd
Southern Water Services Ltd
Thames Water Utilities Ltd
Wessex Water Services plc
Yorkshire Water Services Ltd

Contractors

AMEC p.l.c.
Balfour Beatty Ltd
Henry Boot & Sons PLC
Christiani and Nielson Ltd
Galliford plc
Higgs & Hill Construction Holdings Ltd
Kyle Stewart Design Services Ltd
Laing Technology Group Ltd
Alfred McAlpine Construction Ltd
Miller Civil Engineering Ltd
Edmund Nuttall Limited
Tarmac Construction Ltd
Taylor Woodrow Construction Holdings Ltd
Trafalgar House Technology
George Wimpey PLC

Others

Cementitious Slag Makers Association
Department of the Environment
Department of Transport
Health & Safety Executive
Hong Kong Government Secretariat
H R Wallingford Ltd
Institution of Civil Engineers
London Underground Limited
National Power PLC
National Rivers Authority
Scottish Hydro-Electric plc
Union Railways Limited